MW00448071

VIRTUAL NATIVES

VIRTUAL NATIVES

HOW A NEW GENERATION IS REVOLUTIONIZING
THE **FUTURE OF WORK**, **PLAY**, AND **CULTURE**

CATHERINE D. HENRY AND LESLIE SHANNON

WILEY

Copyright © 2023 by Catherine D. Henry and Leslie Shannon. All rights reserved.

Published by John Wiley & Sons, Inc., Hoboken, New Jersey.
Published simultaneously in Canada.

No part of this publication may be reproduced, stored in a retrieval system, or transmitted in any form or by any means, electronic, mechanical, photocopying, recording, scanning, or otherwise, except as permitted under Section 107 or 108 of the 1976 United States Copyright Act, without either the prior written permission of the Publisher, or authorization through payment of the appropriate per-copy fee to the Copyright Clearance Center, Inc., 222 Rosewood Drive, Danvers, MA 01923, (978) 750-8400, fax (978) 750-4470, or on the web at www.copyright.com. Requests to the Publisher for permission should be addressed to the Permissions Department, John Wiley & Sons, Inc., 111 River Street, Hoboken, NJ 07030, (201) 748-6011, fax (201) 748-6008, or online at http://www.wiley.com/go/permission.

Trademarks: Wiley and the Wiley logo are trademarks or registered trademarks of John Wiley & Sons, Inc. and/or its affiliates, in the United States and other countries, and may not be used without written permission. Virtual Natives is a registered trademark of Catherine D. Henry. All other trademarks are the property of their respective owners. John Wiley & Sons, Inc. is not associated with any product or vendor mentioned in this book.

Limit of Liability/Disclaimer of Warranty: While the publisher and author have used their best efforts in preparing this book, they make no representations or warranties with respect to the accuracy or completeness of the contents of this book and specifically disclaim any implied warranties of merchantability or fitness for a particular purpose. No warranty may be created or extended by sales representatives or written sales materials. The advice and strategies contained herein may not be suitable for your situation. You should consult with a professional where appropriate. Further, readers should be aware that websites listed in this work may have changed or disappeared between when this work was written and when it is read. Neither the publisher nor authors shall be liable for any loss of profit or any other commercial damages, including but not limited to special, incidental, consequential, or other damages.

For general information on our other products and services or for technical support, please contact our Customer Care Department within the United States at (800) 762-2974, outside the United States at (317) 572-3993 or fax (317) 572-4002.

Wiley also publishes its books in a variety of electronic formats. Some content that appears in print may not be available in electronic formats. For more information about Wiley products, visit our web site at www.wiley.com.

Library of Congress Cataloging-in-Publication Data is Available:

ISBN 9781394171354 (Cloth)
ISBN 9781394171361 (ePub)
ISBN 9781394171378 (ePDF)

Cover Art: Carlos Andres Bernal
Cover Design: Paul Mccarthy

SKY10051659_072423

Contents

Preface

Welcome to the Dawn of the Virtual Natives

TODAY, WE FIND ourselves in a period of technological acceleration the likes of which we have not seen since the dawn of the internet itself, and we're greeting it with the same cocktail of awe, inspiration, confusion, cynicism, astonishment, and disbelief. If you've felt this and wondered if you're the only one reacting this way – you're not.

Since the onset of the global pandemic, we have found ourselves caught in a moment that has demanded new solutions for an unprecedented era that has come to affect society permanently. Driven by the necessities of lockdown, multiple virtualized solutions – like using QR codes for tickets, and menus, or having food delivered to your door by people whom you never even see – became the new normal.

But it was more than just using technology to keep us a safe distance apart from each other. Suddenly, and out of nowhere, we found ourselves bombarded from all sides by terms like metaverse, VR, AR, NFTs, and digital assets and collectibles. Crypto, DeFI, Web3, ChatGPT, Midjourney, and DALL-E came to dominate the common lexicon. Technology is suddenly at the core of conversations around the world, an issue of burning intensity.

Blink, blink. What just happened?

We're living through what is arguably one of the most exciting, confusing, and powerful social moments in the history of humanity. The 2020 pandemic drove a great acceleration in the development of multiple forms of technology. The urgent need for "contact free" tech solutions radically changed how we use digital tools to achieve virtual results. We used digital tools to deliver personal virtual "presence" in lieu of face-to-face meetings. We used QR codes to generate virtual menus and expanded digital payment systems to send each other virtual forms of currency. Today, these virtual experiences have multiplied exponentially across industries and businesses, and have come to permeate our lives and our culture. What began as a series of "temporary" adjustments have now become permanent. This Covid-driven shift in technology use has been matched by the simultaneous rise of new, Web3 technologies. The people who are leading both charges are the ones who were already using them before 2020 and today know them best: the Virtual Natives.

This book will introduce you to the Virtual Native cohort and mindset. As builders and creators in these new spaces, we aim to decipher their sociocultural and economic experiences and unpack their expectations of companies looking to engage, market, or employ them.

Whether for work, gaming, or social life, Virtual Natives are driving how we use emerging Web3 and virtual technologies, and evolving culture in the process. They're creating and inhabiting playgrounds for exploration, exchange, connection, and personal expression. And their economic activities are forging a bold new marketplace that is evolving in real time.

Virtual Natives use their devices like appendages to perform multiple tasks, simultaneously. Rather than escape the world through their devices, however, VNs are taking control of their lives by using them to their advantage. They are arming themselves with the tools and knowledge they need to hack their own futures, while discarding old rules, habits, and expectations that no longer serve them. From fax machines and long daily commutes to even standard business hours, Virtual Natives are reassessing their lifestyle, education, and workplaces, and optimizing everything in real time to better suit their individual needs.

This is nothing short of a cultural revolution, and it's happening now.

In writing this book, we have followed in the venerable footsteps of digital anthropologists such as Neil Postman, Marshall McLuhan, Sherry Turkle, and Don Tapscott, all of whom explored the birth of the digital era some 25 years ago.

Now we enter a new period of change, an inflection point with profound technological, cultural, and historical implications, and it's time for reassessment.

Let's go!

A Revolutionary Shift

First there were the Baby Boomers (born 1945–1964), then Generation X (1965–1980), followed by the Millennials (1981–1995), then Gen Z (1996–2009), and the latest group – the Alphas, born 2010 or later. While the level of digital familiarity and expertise has increased with each subsequent group, it's with Gen Z and the Alphas that we begin to identify a cohort, the Virtual Natives (VNs for short), that are using digital tools not just to get the job done, but rather to deconstruct traditional ways of approaching tasks, and to pick and choose only the angles and activities that make sense to them. They're questioning the way things have always been done, going against expectations from previous generations, and using the massive power of computing that has surrounded them their entire lives to rebuild processes and systems to match their own expectations and desires. Virtual Natives are not just the kids who hang out in various virtual realities, as the term might suggest on its surface; they're the kids and young adults who were born into a world where virtualization is increasingly central to their existence, experiences, and expectations.

It sounds like a big shift, and it is. But hold on – are Virtual Natives different from the Digital Natives we've been talking about for the last 25 years? Yes, they are, and in significant ways.

The term "Digital Natives" was originally coined by educator Marc Prensky in 2001.[i] He used the phrase to describe the late Millennial generation who "are used to the instantaneity of hypertext, downloaded music, phones in their pockets, a library on their laptops, beamed messages and instant messaging." These were kids who laughed at their elders who printed out their emails and bought books from a brick-and-mortar bookstore. Instead, they were happy to leave their emails online and order their books from a little (at the time) company named after a big river. They spoke "technology" fluently and adapted to digital formats quickly and almost effortlessly. By the time they were college graduates, Prensky said, they had spent "less than 5,000 hours of their lives reading, but over 10,000 hours playing video games" and more than 20,000 hours watching TV, which was their main cultural influence.

This was all a big shift from what had gone before. Prensky's accurate description and catchy title caught on and entered general usage.

While the Digital Natives moved happily and effortlessly in the newly digitized world, the online world of 2001 was still very much a digitized version of the original analog world. Almost all the online functions that anyone could perform back in 2001 were computerized versions of existing physical-world processes that replicated the original process but did little to change or improve it. Most of those were one-way transfers of information, or transactions, made slightly faster and easier by virtue of being accessible through a computer. For example, annual reports were no longer printed but posted on corporate websites as PDFs, still looking exactly as they had when they were mailed out in envelopes. College students back then may have played video games for more hours than they read books, but they still did read the books that they ordered on Amazon, and watch broadcast TV. Digital Natives no longer asked each other for turn-by-turn directions, but got them from Google Maps on their computer at home, and then printed them out to take along with them in the car.

What has changed, and made the new Virtual Native cohort possible, is that technology has advanced to the point that it is now possible to achieve many goals in ways that no longer bear any resemblance to the original non-digital process that preceded them. Where a pre–Digital Native would have sent a physical letter, and a Digital Native of the early 2000s would have sent an email whose format still mimicked that of a business letter (and was still called "mail" and would land in the recipient's "inbox"), a Virtual Native has abandoned all links with the postal past and will have a video call, send an emoji-laden What's App message, or will create and post a video and tag their friends. The objective of sending another person a message has been reached by all three methods; but the Virtual Native approach is so far removed from the original process that Virtual Natives today likely have no idea or interest in posting an envelope.

This is the heart of the difference between Digital Natives and Virtual Natives. Both groups use digital tools, but Digital Natives tended to use those tools in a way that replicated the physical process that preceded it. If you, dear reader, are a Digital Native, chances are high that you're still using email, and probably using it more at work than you use group chat tools like Slack or Yammer. You probably haven't realized that what you're doing is just a computerized version of the previous process of writing on a piece of paper and sending it to your colleague via an interoffice envelope. Virtual Natives, on the other hand, completely lack the experience of the original, pre-digitized past, and are free to entirely reimagine how things can be done, using the extremely powerful digital tools at their disposal. Why waste time sending one message to one person once via email when it's much more efficient to post a message in the group Slack, Discord, or Teams channel, which then automatically becomes part of the shared history of the project?

Virtual Natives swipe instead of typing, use YouTube and TikTok to search instead of Google, and are often more comfortable meeting new people as an avatar in an immersive 3D environment such as *Fortnite* or Rec Room than they are with dealing with others in person. To VNs, physicality is less important than the end result: the feeling of live connectedness with others. This is a whole new generation with new tools and new habits, and those in turn are driving new ambitions and expectations.

New ways of thinking give rise to new economies. Just like Digital Natives and their excitement at the dawn of the dot-com era, Virtual Natives are eager to leverage the burgeoning Web3 economy and reap its many bounties, in an attempt to become masters of their own economic destiny. But Virtual Natives will go further than the Digital Natives could, thanks to the improved power of the tools in their hands. Generative AI, for example, will level the playing field so that young bedroom entrepreneurs with no formal education and on a shoestring budget can take on big business and compete with brands in platforms where the physical properties of a product – e.g. a thick, cashmere sweater – no longer has meaning, and must instead be re-imagined and promoted as an emotion. All of these movements have seismic implications for the future of business, entertainment, and culture.

In this book, we explore how different this generation is to its predecessors in unique and significant ways. As we studied Virtual Natives and what sets them apart, we identified 10 main themes that, together, define the unique Virtual Native mindset and approach to life, the universe, and everything. We cover these in the first half of the book, along with the VNs' unique habits, behaviors, and thinking patterns that led us to identify each theme. We'll also look at the awesome technical powers they have grown up with that have necessarily shaped their activities, expectations, and worldview.

In the second half of the book, we explore the implications that the VNs' powerful cocktail of new technologies, habits, and expectations will have for the future of work, play, education, entertainment, culture, and social life. We dive into the potential challenges that will arise for people trying to understand how Virtual Natives' innate virtualization has already impacted their social networks, values, behaviors, economic prospects, and expectations and, importantly, how VNs will balk when confronted with anything that doesn't instinctively feel right.

In both halves, we'll look at examples of how VNs are rethinking the world and evolving it on their own terms. We'll also look back at other moments of significant technological and cultural disruption for parallels and insights that are relevant to understanding what is happening in today's new era of virtualization.

We spoke to many people in the Gen Z and Alpha contingent as part of our research. Because the majority of our interviewees are children, we have left their last names out of our reporting. Any quotation that is listed as coming from someone with a first name only, and which is lacking an endnote, is the result of a direct interview.

The lesson to learn from Virtual Natives is not that they are inaccessibly different and eternally "other," but that they are the product of a new paradigm, an intersection of time, place, and tools. This is an important nexus among technology, media, and culture not seen since the dawn of the internet itself. Those of us who were born in any other generation can take inspiration from the discoveries of VNs both to understand them better in the home and workplace, and perhaps to discover new efficiencies and ways of thinking that can bring welcome change to our own workplaces, professions, and personal lives.

Twentieth-century futurist Alvin Toffler famously said, "The illiterate of the 21st century will not be those who cannot read and write, but those who cannot learn, unlearn, and relearn."[ii] The wheel is turning, the times are changing, and once again, it is time for all of us to learn, unlearn, and relearn, most particularly from those who have the most to teach – the Virtual Natives. Today, watching the sun rise on the 3D internet, we can only wonder what marvels they will create.

Catherine D. Henry
Leslie Shannon
September 2023

PART

1

A Generation Unlike Any Other

IN OUR ASSESSMENT of the preferences and activities of Virtual Natives (VNs), we've found 10 recurring themes that, taken together, express the core of how VNs differ from previous generations. Some of their behaviors are the product of unique historical factors, such as the Covid-19 pandemic and the 2008 financial crash, but most of them flow from the fantastical abilities to communicate and create that Virtual Natives have been granted by the digital tools readily available on the screens that surround them.

We'll start by examining each theme individually in dedicated chapters, then, in Part Two, zoom out to look at how the themes work together to build the Virtual Native worldview – and why we older generations may ultimately end up adapting to them, instead of the other way around.

1

Virtual Is the New Digital

You may be familiar with the video of a one-year-old baby who, before she could talk, had figured out how to turn on an iPad and open her favorite pictures and apps. This isn't so unusual; since the iPhone's 2007 arrival and the iPad's launch in 2010, harried parents have become much more likely to hand their toddlers a smartphone or tablet as a distraction instead of a rattle or a set of jingling keys. So what does this baby do when she's given a magazine? She tries pressing on the images. They don't move. She turns the pages and tries swiping. No response. Puzzled, she tests her index finger on her thigh; is it pressing right? Is her finger working? After a few leg pokes, she decides that it *is* working and tries it on the magazine again – no luck! Now anxious, the baby begins to squeal. Why won't these images open? What is *wrong* with this paper thing, why won't it work the way it's supposed to?[1]

Similarly, there are multiple videos out there of slightly older children, trying to swipe the screen of the television set to change the channel, and becoming equally frustrated when the device doesn't respond as they expect. These are Virtual Natives.

Virtual Natives (VNs) are the people who have known nothing their entire lives but fully digitized versions of what were originally analog activities. This experience is the foundation of not only how they perceive the world, but also how they are shaping it as they grow to reflect their own lived reality. Digitization has led to VNs being able to "virtualize" their experiences, or, that is, to select only what they see as the best and most useful elements of what were previously entire processes, while discarding elements that seem inconvenient, wasteful, or useless. We'll look at examples of how this plays out in practice in this chapter, and come back to explore the larger ramifications of this mind shift in Part Two.

By 2025, there will be more than two billion Generation Alphas,[2] the largest generation in history, who have been born into more diversity and more technology than ever before.

The Alphas, like many of the Gen Zs that precede them, have never known a period when smartphones, apps, and video calls did not exist. As babies, they learned that their media is interactive, and when it's not, they consider it to be broken. In fact, this is a crucial point for understanding how Virtual Natives differ from others: How do they identify what is, to them, broken, and how do they use the digital tools they've handled since birth to create a more desirable fix? To create these fixes, VNs are learning and teaching each other new codes, new ways of connecting, and hacking life in a radically different way than previous generations.

Let's start with some definitions. We define Virtual Natives as the people, largely belonging to the Gen Z and Alpha cohorts (though not exclusively – it's a mindset as much as it is an age group), who are using digital tools to craft a life that makes sense to them, on their own terms, rather than blindly following the patterns laid down by those who have gone before them. Virtual Natives are not just the kids who spend a lot of time in virtual reality, as the term may casually indicate; they're the kids and teens with the courage and the skills to deconstruct the world as it is, and to rebuild it for themselves as they'd prefer it to be.

Merriam Webster defines the word "virtual" as "being on or simulated on a computer or computer network" or "occurring or existing primarily online."[3] For the purpose of this book, we define "virtuality" as the grand sweep of digitized experiences and functions that are evolving along with new technology, enabling Virtual Natives to express themselves in all facets of their lives. Their education, play, work, and collaboration are all seamlessly both digital and physical and, to them, indistinguishable, with neither format being materially "better" or "worse" than the other. Technological tools and devices are to Virtual Natives seemingly natural extensions of their being that allow them to reach their desired results without worrying over the formalities of "how," focusing instead on the "why" and, more important, "what's in it for me." And that's exactly what Virtual Natives do: refuse to accept traditional processes that don't make sense to them, using their powerful digital toolbox to reach target outcomes. We'll see many examples of this virtualization dynamic in the coming chapters.

Ziad Ahmed, the 20-something CEO of JUV Consulting, who specializes in helping companies understand the Gen Z cohort, says it directly: "We're not going to do something just because that's the way it's always been done."[4]

Finding Your Fam

While a key part of VNs' development lies in their familiarity with technology from birth, it's also undeniable that the global pandemic of 2020 played a formative role in the construction of the Virtual Native worldview.

As Covid began to grip the world, and country after country reacted with strict lockdowns and the prohibition of public gatherings, people around the world suddenly found themselves unable to attend events with landmark life significance, such as graduations, weddings, even funerals. Older generations grudgingly moved their meetings and happy hours with friends to platforms such as Zoom, where they could see each other as they raised their wine glasses.

But not the Virtual Natives. They went into their games.

Kindergarten and grade school graduations took place in Minecraft. Live video calls with celebrities streamed on TikTok. Concerts were held in *Fortnite*, and funerals and other memorial services took place in *Animal Crossing*. In fact, *Animal Crossing*, which had the fortune of launching its strongly community-focused New Horizons variant just as the Covid lockdown began, became a common location for meaningful gatherings of all kinds, its gentle avatars helping participants to share warm and loving feelings during tough times. And Virtual Natives were already there.

Not allowed to leave the house for Halloween? Do your trick-or-treating in *Animal Crossing*! Unable to get together in person with your friends from school? Join them online in endless games of *Among Us* or *Rocket League*! Miss seeing and sharing the latest fashion trends? Kit out your digital avatar with a hot new look and take plenty of selfies in ZEPETO!

Organized religious services, which moved online early, revealed a difference in expectations around participation between their older congregants and Virtual Natives. "We've seen a hunger in Gen Z for more experiential stuff – something they get to participate in rather than receive. They want to belong to a community rather than an audience," commented nondenominational Pastor Zach Lambert.[5] And for those with no interest in religious services? Immersive yoga, kickboxing, and meditation apps abound!

At a time when a multiplicity of one-way broadcasts from both school and local communities left kids feeling alone and lonely, digital platforms were able to provide Virtual Natives with much-needed engagement, community, and a sense of belonging.

Like babies swiping iPads, Virtual Natives intuitively use digital spaces to meet their friends or watch YouTubers from other parts of the world, and turn to virtual worlds as places to connect with wider, meaningful communities. A 2022 study found that players of Roblox, *Fortnite*, and Minecraft all spent more time in those worlds – and spent more money there – when they also met their friends and did their socializing there.[6] Whereas Millennials have been known to overshare, documenting the most minute details of every meal, moment, and milestone, in often overly glossy and perfected tableaus, Virtual Natives see social media much more as a tool for understanding the world and are more interested in creating than curating and cataloging their lives.

Joining communities online, finding their "fam," or friend family, helps Virtual Natives answer that very human need of wanting to feel part of something larger, of having a sense of solace and purpose, and belonging. When asked in 2023, a majority of teenage girls in one survey confirmed that social media has had a positive effect on their lives, by connecting them with others, helping them find friends outside their immediate geographies, and giving them information about mental health and other resources that they couldn't find elsewhere.[7]

Mike Schmid, head of publishing for Rec Room, sees online worlds as an extension of daily life, saying, "We don't use the term 'metaverse.' We don't talk about ourselves in the sort of creepy way that some other experiences do. We're just supplementing real life."[8]

The implication here is that for Virtual Natives, both digital spaces and avatars are worthy stand-ins not just for gameplay, but also for humanity's significant rites of passage. Digital worlds are valid places for both community and communion, and VNs feel that actions that are taken there (weddings, graduations, funerals) have meaningful significance in the physical world.

Hanging out is a rite of passage, and each generation has had their "third space" – not school or work, and not home – where they hang out and socialize. Boomers met up at diners and soda fountains, Gen X in record stores and malls, and Millennials in Starbucks. For Virtual Natives, digital spaces are not just meeting spaces, they're also community spaces in which they can truly connect with others. This may be especially meaningful for VNs, many of whom may never know what it's like to have a community at work. In this case, these digital "third" spaces may become even more important. You can be a genuine human participant in a group and its activities, seeing, being seen, and interacting with others, without ever meeting anyone in person.

The Road to Virtualization

Virtual Natives were practically born with digital devices in their hands, and Covid accelerated their embrace of digital spaces as a practical solution to a real-world problem. But the pandemic was (we hope) a one-off, a black swan event that isn't a persistent motivator for long-term behavioral change. There must be another factor at work that has driven Virtual Natives to reject the status quo and to seek different, more virtual ways of living their lives.

Virtual Natives see Millennials as a cautionary tale as a massive cautionary tale. The older members of this group have had terrible luck with their timing. First, they had to deal with the Great Recession of 2008, when 17.2% of people aged 20–24 were out of work, and in the decade since then, they've faced higher levels of student debt and more stagnant wages than the generations before them. In 2019, it was calculated that Millennials earn 20% less than the Boomers did at the same age, a devastating drop that affects everything from daily anxiety levels to their ultimate ability to retire – *ever*. This is such a looming and ominous reality for younger generations that we'll return to this topic again; it's a driver for multiple behaviors among the VN cohort.

As Millennial Charles Bryant put it when interviewed by *Fortune* magazine in 2022, "I followed the rules that they gave us. I played the game and went to school and I did very well. Every job I've worked, I've worked very hard and it didn't pay off. It feels very unfair."[9]

Exactly. It is unfair. And the Virtual Natives are highly aware of this dynamic; it's a conscious motivation for them. In a March 2023 Twitter thread responding to a poll of older generations that found Gen Z workers to be unmotivated, one VN retorted, "Lack of effort/motivation/productivity/drive: We don't care about the job, it's something we have to do, not want to do. The economy sucks, and we have no guarantee of stable CAREER future. Guess what? Boomers, Gen X, Millennials: It's YOUR fault."[10]

"Working for the man" and playing by his rules didn't do the Millennials any good, so it should come as no surprise that Gen Z and Alphas are looking for other paths toward the life that they want to have. Digital tools, platforms, and services give them those paths, potentially restoring the life control that they perceive the Millennials as having lost.

But there's more to it than just using a set of tools. At a conference in late 2022, Lisa Costa of the United States Space Force spoke about how NASA finds some of their best workers among their youngest hires. It's not just that Virtual Natives are very comfortable with using computers; even more important is what she described as their "cognitive training." NASA has found that VNs are very good at prioritizing the important and deprioritizing, even discarding, the unimportant, making them uniquely able to make rapid and confident decisions based on partial evidence. This cognitive training comes not only from fast-moving gaming environments, but also from the TikTok world, in which content creators are used to grabbing viewers' attention within the 15-second time frame that was TikTok's original video limit. It's this ability to discern what's important, often at speed, sort the wheat from the chaff, and focus only on the things that deliver value, that is the larger skill that sets VNs apart from older generations.[11]

The Essence of Virtualization

We began our examination of Virtual Natives with this definition of "the virtual" because it is so critical to understanding the VN approach to everything. To them, "virtualization" doesn't mean just doing the same thing as before, only digitally; it means tearing key constructs down into components and only retaining those that produce results or provide meaning. The focus is on *outcomes*, not on processes, no matter how expected or revered those processes may be.

This approach to the world may seem heretical and unthinkable to older cohorts, but ultimately, Virtual Natives will cause us all to rethink how we approach life, work, and leisure, and how to locate the true value at the center of many processes that the rest of us may not have thought about analytically at all. In virtualization lies power.

> The young do not know enough to be prudent, and therefore they attempt the impossible – and achieve it, generation after generation.[12]
>
> —*Pearl S. Buck*

2

What Is Reality, Anyway?

WHERE OLDER GENERATIONS may make a distinction between digital worlds and the "real" world, meaning the physical world, VNs are much more likely to consider what happens in digital arenas to be *just as real* as that which happens in the physical world. After all, it's on digital platforms that they do most of their communication with friends and learn about the world. Whether a VN is talking to their friend in person or as an avatar in a virtual space, the relationship and the emotions are real, and that's what matters.

In 2021, ahead of the release of *The Matrix Resurrections*, which was coming out a full 22 years after the first appearance of *The Matrix* in 1999, star Keanu Reeves found himself explaining the plot to a group of teenagers who had been born after the original movie was made.

He outlined the basic plot, then got to the twist at the center of the *Matrix* franchise. "There's this guy who's in a kind of virtual world. And he finds out that there's a real world, and he's really questioning what's real and what's not real."

For those of us who first saw the *Matrix* movies when they came out, we remember how horrifying and deeply unsettling it was to discover that everything the characters Neo and Trinity were experiencing as "the real world" was a simulation, an elaborate digital trick pulled on their sensory system by forces beyond their control.

Instead of being blown away, one girl just looked at him and asked him, "Why?" Reeves was unsettled. "I was like, 'What do you mean?' And she was like, 'Who cares if it's real?'"[1]

Virtual Natives have been so surrounded by digital representations of, well, everything all their lives that they have developed a view of reality that blurs the line between the physical and digital worlds.

Instead of seeing that "reality" lies only on the physical side of the line, as previous generations may tend to do, their definition of reality is that of a continuum, which encompasses both the physical world and the myriad digital spaces that they have participated in all their lives. Digital, physical – they're both real, and the experiences that VNs have in both are real, and the relationships that they form in both are real, and the places that they go in both are real. And it's just not a big deal. Sorry, Keanu.

The Time and Space Discontinuum

Mason was in eighth grade. Bored during his class, he turned to his friend Sunil and said, "Hey, wanna watch a movie together after school?" "Sure," Sunil said, "do you have Netflix?" Mason responded, "Yeah, meet you after school at 3:30, and we'll pick something out then, okay?"

So far, so normal. But what was actually happening is probably not the scenario you were expecting. The year was 2021, and Mason and Sunil were each at home in their own rooms, attending their classes online during distance learning. Mason and Sunil – who don't even go to the same school – each had their browsers open to their current class, as they were supposed to, but also had Discord open, in which they were chatting to each other whenever the educational side of things got a little slow.

They were effectively passing notes to each other in class. It's just that in the 2021 version of passing notes, they were speaking out loud, not writing (making sure they were on mute in their school window, of course), and they were going to different classes at different schools at the time.

When they met up after school to watch their chosen movie, Sunil didn't come over to Mason's house, but instead they met up again in Discord. Mason has Netflix access, which Sunil doesn't, so after they chose their movie, Mason started playing it, then shared his screen in Discord with Sunil. They were able to watch the show together, talking and laughing with each other, just as if they were sitting next to each other on a sofa in the same living room. They even had popcorn.

This scenario, which Mason played out almost every day of his distance learning, was the main reason that Mason was deeply disappointed when in-person learning resumed as he entered ninth grade. Even though he was excited about going to high school, and *physically going* there, he was quite bummed that "I won't be able to talk to Sunil during my classes anymore. Going to school won't be as much fun without my best friend." Mason didn't see the resumption of physical schooling as a superior way of connecting to other people when compared with his digital afternoons with Sunil; for the two boys, their time spent together in Discord had just as much value and validity as time spent physically together would have held.

Happily, once in-person school resumed, Mason and Sunil kept up their afterschool Discord meetings – sometimes watching a movie, usually gaming, always talking to each other as they did – and Mason found other friends to pass physical notes to in class. But Mason and Sunil are far from unique. In a 2022 study, Gen Z gamers were found to spend roughly twice as much time hanging out in social game spaces with their friends as they spent with their friends in the physical world.[2]

For Virtual Natives, actual distance means literally nothing. A recent survey showed that under-18s were twice as likely than those over 35 to enter MMOs (Massively Multiplayer Online games) such as Roblox or *Fortnite* with the sole purpose of socializing, rather than playing the game.[3] In fact, according to Everyrealm, Generation Alpha is the first to spend more time in social video games than on social media.[4] Whether you're together in person or on a screen just doesn't matter; the point is to spend time together, and both methods accomplish that.

But there are plenty of people who *are* in there to play the game. Mason's older brother, Marshall, is on a *Fortnite* team whose other members he's never met in person. They coordinate their strategies with each other by talking on Discord during games, and it usually gets pretty exciting. "No!! Don't you see the guy on the tower? TAKE HIM OUT, THEN MOVE LEFT! GO GO GO GO!" – and so on.

They've evolved into a tight-knit, finely tuned winning team, entirely through speaking to each other as they play. When asked if it was important to see the actual faces of his teammates during games, Marshall responded, "No, we don't turn on the video camera to see each other during a game – we can all see our avatars, and running a video camera would take processing power away from the game."

Traveling with Your Mind

In a world of YouTube and TikTok, the way to communicate with others at a large distance (and therefore an inconvenient time difference) is easy – just post a video that others can watch later. In the videos posted by Virtual Natives all over the world, viewable at any time and in any place by any other Virtual Native anywhere else in the world, VNs are sharing their hopes, dreams, accents, favorite foods, favorite clothes, new makeup styles, and everything else imaginable, in one giant global mashup of asynchronous communication. Both distance and time zone are annihilated in a single blow, and everything – EVERYTHING – becomes shareable and accessible across multiple platforms.

When the first bubble tea shop opened in his neighborhood, Marshall already knew what bubble tea was – not by hearing about it from his friends in California, but by seeing videos about bubble tea made by kids in Hong Kong. (And enough of his friends also know what bubble tea is, for the same reason.) Mason can imitate remarkably passable Australian and British accents – and can tell the difference between them – because he's been hearing them online for almost his entire life. VNs don't need to have done any actual travel to be aware of – and to appreciate the worldview of – people from very different places, whose life experiences are of very different things.

Marshall's favorite pastime, aside from playing *Fortnite* and drinking bubble tea, is watching Premier League football from the UK, which he got hooked on after playing many years of *FIFA* on his Xbox. None of his other friends in California are fans, so while Marshall watches live games on TV at home, he also streams live YouTube channels on his phone of groups of fans in the UK who are also watching the game in real time. When he cheers, they're cheering, and when he's groaning at a dumb move, so are they. This reinforces Marshall's fandom, and gives him a feeling of shared community and passion, making every game an event that connects him to others who share his interests, even if only one-way and from afar.

Deion is one of the Virtual Natives whose entire life trajectory was changed by being exposed to other cultures online as a teenager. A massive fan of Japanese manga, he began playing Japanese video games as a young teenager to more fully immerse himself in the world he saw in his graphic novels. This led him to play online role-playing games, or RPGs, such as *Final Fantasy*, that had been created in Japan. But playing in English just wasn't giving him the full Japanese experience, so he started teaching himself Japanese in his spare time and playing RPGs through their Japanese-language portals.

Deion found that within these digital worlds, his rudimentary ability to express himself in Japanese was welcomed, and he soon made real friendships with Japanese RPGers his own age. By the time he graduated from college, Deion was fluent enough in Japanese that he ended up actually moving to Japan to become a game designer and animator. Thanks to the awesome global reach given to him by his digital tools, Deion was able to spend his teenage years physically in California, while being mentally in Japan as often as possible. His world became simultaneously larger and smaller.

For those who are deeply curious about other cultures but are not willing to go the extra mile forged by Deion, language has been the one remaining obstacle that separates VNs of one part of the world from another. But with real-time translation improving on an almost daily basis, it's only a matter of time before that final barrier also falls.

The message here for the rest of us is that Virtual Natives have an extremely different perception of what time and distance mean, and how they can be conquered with validity by digital channels, than older generations do. And this elimination of both time and distance as barriers to effective communication, both with friends and strangers, means that Virtual Natives have a profoundly more integrated view of the globe than any previous generation has ever had.

What this will mean for the future is an open question. But it is likely that twentieth-century divisions of "people like me" and "people not like me" will erode. For example, in January 2022 Jordanian teenager Issam Alnajjar broke TikTok with his cheerful, catchy rendition of his Arabic-language song "Hadal Ahbek," which led to his getting noticed by music industry executives in North America.[5] The song was one of the most-streamed on Spotify in markets as diverse as the United States, UK, Saudi Arabia, Malaysia, and India, and symbolizes the effortlessly global nature of Virtual Native culture in the modern era.

Another symbol of the global universality of Virtual Natives is Puerto Rican reggaeton superstar Bad Bunny. In 2023, he was Spotify's most-streamed artist for the third year in a row, an even greater accomplishment when you realize that he sings not in English, but Caribbean Spanish. American record industry executives had maintained for years that only songs in English could be hits in the United States, but, with digital access to global acts like Bad Bunny and of course the entire rich world of K-pop, sung in Korean, fans are proving them wrong. Who cares if you can't understand the lyrics, as long as the music is good? As the New York Times put it, "[Bad Bunny's] recognition at the Grammys, the Billboard Music Awards and the Video Music Awards is not a result of his entrance into the mainstream but rather of the mainstream being forced to reckon with the purchasing power of his legion of fans."[6] And these fans found him, and plenty of other international music groups that they have fallen in love with, on digital platforms.

Another major content producer, Mr. Beast, who has over 130 million followers on YouTube, has started dubbing his English-language videos into 11 different languages, including Hindi, Thai, Spanish, Indonesian, and Arabic. "It's important not just to reach the people who speak English [with your videos], but to reach everyone. If you do the top 15 languages, you can reach basically 90% of the world."[7] Virtual Natives don't see their potential market as limited by geography or language – with digital assistance, their market is the entire planet.

This use of digital media and platforms to connect to other peoples and cultures physically located in other parts of the world is part of why Virtual Natives don't draw a fine line between physical reality and digital reality. They've spent their whole lives communicating both in real time and asynchronously with people physically located elsewhere, whether they're around the corner or around the world, and they've learned that there's just not that much difference between talking to someone in person and talking to them on a screen.

Everywhere and Anywhere

In a world of FaceTime, TikTok, YouTube, and multiplayer gaming, it's not a surprise that Virtual Natives are accustomed to interacting and forming relationships with others at a distance through digital means. But what about places?

The year was 2017. Teri and her family had moved to California two years before, and they were still exploring the different adventures to be had in their large and varied state.

This weekend, they were down in Los Angeles, five hours south of their home in San Jose. Both Teri and her husband had been to Los Angeles before, but Jade, 12, and Gabriel, 10, had not. That's why she was so surprised when, as they drove along the shoreline in Santa Monica, she suddenly heard both children say from the back seat:

"Mom, I *know* this place. We've been here before, right?"

"Yeah, I remember it, too! See that funny-shaped restaurant on our left? There's going to be a gas station right after it."

There was.

"Oh, and then there'll be a kind of bridge that people can walk over to get to the beach! See, there it is!"

A pedestrian bridge, which hadn't been visible when Jade and Gabriel first predicted it, loomed into view, to much delighted laughter from the back seat.

"Wait, you've never been in Los Angeles before!" said Teri, stunned. "How do both of you seem to know this neighborhood?"

Suddenly the light bulb went off in both heads at once. Jade and Gabriel looked at each other, and yelled, in unison, "*Grand Theft Auto!*" Jade then added, "That's totally what it is. I've robbed that gas station tons of times!"

The video game *Grand Theft Auto V*, which the kids knew from hours of play on the Xbox at home, is set in the fictional city of San Andreas, which is a mashup of various cities, including the Los Santos neighborhood, which is based on Los Angeles. It turns out to be *very* closely based on parts of Los Angeles, copying 3D versions of existing structures so faithfully that Teri's children were able to recognize a string of buildings as they drove through Santa Monica.

In our discussion of time and space in virtual worlds, we've looked at how time and space are collapsed during interactions with other people, and how direct connection with others is simultaneously turning Virtual Natives into citizens of the world, and enabling them to become contributing members of meaningful communities. As the *Grand Theft Auto* Los Santos/Los Angeles experience shows, even gaming on a 2D screen can give players such a strong sense of place that they literally feel like they've *been there before* when presented with the same structures and layout in their actual physical surroundings.

And this is the point. While Jade and Gabriel had never been to Santa Monica before physically, they had been to a digital place that exactly copied Santa Monica, and their memories of the place are real. The memories created while in digital realms (as well as in the physical world) are the kind of memory known as *episodic memory*, memory generated by experience. This is a richer, more easily recalled memory than *semantic memory*, which consists of information, facts, and concepts – the kind of knowledge you get by straightforwardly memorizing something. As the author David Rose explains in *SuperSight*, his book about augmented reality, "The brain is adapted to absorb and combine visual, aural, and other sensory context cues, and to use any or all of these as hooks for future recall. We remember more effectively when we encounter knowledge in rich, real-life contexts."[8]

Studies have borne this out: Memories generated in fully digital worlds are indeed likely to be confused with memories from the physical world, and vice versa, even when the physical world memory includes tactile elements currently not available in screen-based experiences.[9]

Let's take this one step further. As we mentioned earlier, older people speaking of the digital world tend to contrast it with the "real world," as though there is something inherently "unreal" about digital experiences. The experience of Jade and Gabriel demonstrates that this is not an accurate division. To the user, the experiences in both physical and digital realms are real. And so are the memories, and the emotions. For Virtual Natives, reality isn't firmly anchored in the physical world, and only dimly mirrored in the digital world; reality, what is real, happens in both the physical and the digital worlds, and VNs effortlessly combine the two during almost every moment of their day.

If you're a parent, you can test this yourself. First, try asking your Virtual Native, "How was school today?" Chances are high that you'll get the universal answer, which is, "Fine," delivered with a shrug. Then try, "What did you do online after school today? Play any good games?" This question is much more likely to generate an excited retelling of a lively experience, plus maybe stories about some dumb moves made by their friends that made everyone laugh.

Doing Everything, and All At Once

Everly is at her desk in her bedroom in Minneapolis, doing her homework. She has two screens on her desk. On the screen directly in front of her is her current game of Minecraft, where she's finishing up a build. On the screen to her left is Discord, where she's watching her best friend, Olivia, navigate a dystopian future as an alley cat in the game *Stray*. They're each playing their own games, while talking to each other and commenting when the other player does something noteworthy or cool. Occasionally, Everly's phone will ping as well, with a message sent by another friend who's asking for opinions about new clothes choices. Everly shifts her attention from one screen to the other as needed, smoothly, effortlessly, and completely at home swimming in multiple streams simultaneously.

One additional feature of the smooth continuum between the physical world and digital representations of people, places, and things experienced by VNs is that they combine them all, all the time. Even while doing activities in the physical world – walking the dog, brushing their teeth – Virtual Natives are likely to be sharing their attention with something digital, offered by some screen in their vicinity. And it's this ubiquity, this near-continual mashup of the physical and the digital, that makes the teenagers that Keanu Reeves spoke to wonder why his *Matrix* character was so fussed about trying to nail down what the "real" world is. Reality is just all of the above, happening all the time. Everything, everywhere, and all at once.

3

Fluid Identities

VNs HAVE AMPLE opportunity to explore individual elements of who they are and how they represent themselves to the world, both online and off. Far from the narrowly gendered, racial, or social strata of previous generations, this is the most diverse generation so far, and they choose to define themselves in their own terms. Online, they can highlight or explore certain interests or aspects of themselves while downplaying others, and they can experiment, experiment, experiment.

Having grown up with smartphones and their apps since infancy, Virtual Natives have had a camera within arm's reach for pretty much their whole lives. And from the start, VNs have been turning that camera on themselves as much as they've been capturing the world around them. The word "selfie" was first used by an anonymous Australian on an online forum in 2002,[1] and by 2013, selfies had become enough of a thing that the *Oxford English Dictionary* named the term its word of the year.[2]

Then, in 2015, everything changed when Snapchat introduced its first Snapchat Lenses. These used active facial mapping to apply digital effects to the user's face in real time. In the early days, it was all rainbow barf and kitten ears, but it wasn't long before lenses became more subtle and started thickening eyelashes, enhancing cheekbones, and generally making users of all ages look pretty darn good, without necessarily flagging that a filter had been used.

While digital lenses and filters (lenses are applied to an image in real time and are visible as you snap the picture; filters are visual changes applied after a photograph has been taken) are used by millions of people of all ages every day, it is the Virtual Natives who have always had the capability of shape-shifting, of looking like something other than the physical body they inhabit, for as long as they've had consciousness. Many have thousands of selfies, but never an actual, physical photo of themselves. Combine the ability to change how you look in every photograph that you take, always, with the ability to choose and customize any

27

kind of avatar you like in any game you play, always, and we find that Virtual Natives have developed a very fluid sense of their own identity. They not only experiment with how they look, but also with who they are, and who they can be.

"My avatar is who I want to be on that day," says influencer Monica Quin on South Korea's hugely popular metaverse social media app, ZEPETO.[3] "In the real world, it's not easy for us to cut our hair and then grow it back, but in the digital world, we can do that with just one click."

VN experimentation with different ways of being isn't limited to avatar use. Shawn Whiting, head of creators for virtual meeting space Rec Room, relates that "when I ask a new creator to show me the work that best represents their personality, they'll often come back with something like four separate YouTube channels and two different Twitter accounts, each showing a different side of themselves. When I ask them which one is the real them, the answer is usually, 'I'm all of these people!'"[4]

Online personas offer a way for anyone to create versions of themselves that reflect their personality and interests online, enabling them to explore multiple ways of being in the world. Just as Virtual Natives use their digital tools to deconstruct and reconsider processes like watching movies together and what it means to "be somewhere," they are also using the tools available to them to deconstruct and reconsider their very identity – or at least the identity that they present to the world.

Skins, Brands, and Identity

Sometimes the identity fluidity of VNs serves a very practical reason.

Emiliano, 15, was playing *Fortnite*. As his mom passed by his bedroom door, she looked at his screen and asked, "Which one are you?" Without looking away from his gameplay, he sighed audibly. "I'm the one on the left, in red."

"Wait – you're the one in *red*?" she asked.

"Yes!" he replied, visibly annoyed.

"The girl?"

"Of *course*," he blurted. "*Everyone* plays as a girl!"

What he did not mention then (probably because it seemed so painfully obvious), but explained later, is that from what he sees, perhaps 9 out of 10 *Fortnite* gamers of all genders, including those at pro levels, play as female avatars because those figures are physically slighter – and therefore harder to hit. "You know, their waists are a lot smaller, which makes their hit box smaller, too."

"So you're playing as a girl purely as a strategic choice?"

"Well, yeah. Their avatars are also a lot more stylish, too, so it's fun."

As strategic a gaming choice as this seems to be, we should not lose sight of the fact that for many male gamers of earlier generations, playing as a female would have been simply unthinkable. It is significant that VNs find selecting nonrepresentative avatars just, well, logical, if it solves a problem. No big deal – which is why Emiliano found it weird that his mother even thought his avatar choice was worthy of discussion.

What does it mean for a whole generation of young men to see themselves represented as stylish, powerful females, literally kicking each other's asses? And cheering for these female warriors during games? When taking on other "skins," or identities, is as common as changing shoes in the physical world, the willingness of VNs to perceive themselves in bodies, genders, and ethnicities far different from those they were born with necessarily increases.

Unlike the physical world, the "skin" you inhabit online does not limit your experience; rather, in virtual worlds it liberates our capacity for self-expression. With the ability to fluidly move from one community or world to the next, one's identity is equally fluid. And this matters, because VNs are born into a world in which old labels and categories are becoming increasingly blurred.

Love Is Love

Virtual Natives have been born into multiethnic and blended households, increasingly with families that included parents of the same sex. In 2022 there were 1.3 million same-sex couple households in the United States[5] – more than double the number recorded in 2008 – and about 60% of them were married. Popular, long-running television shows like *Will and Grace*, *Modern Family*, and *Grey's Anatomy* portray loving mixed and gay marriages as part of the American fabric. The term "Dads" and "Moms" is used frequently in media.

Same-sex marriages are now legal, and normalized. A survey by Pew Research showed that over half of Gen Z believes that same-sex marriage makes society better.[6] Fifth-graders in the conservative state of Indiana surprised many in 2023 when one class led a school walkout as a protest against an anti-LGBTQ+ bill then under consideration by the state legislature. As one 10-year old put it, "We are trying to protest to make sure that kids have rights and won't be stopped from expressing their whole self."[7] Overall, just under 50% of Generation Z identifies as "completely heterosexual," compared to 65% of Millennials.

"I'm Whatever, and It's . . . Whatever"

At 18, Lily-Rose Depp, daughter of actors Vanessa Paradis and Johnny Depp, said, "People don't have to label themselves and say, 'I'm straight,' or 'I'm gay,' or 'I'm whatever.' If you like something one day then you do, and if you like something else the other day, it's whatever. Kids don't need to label their sexualities. It's not that big of a deal."[8]

Lily-Rose isn't alone in feeling that naming their sexuality isn't a big deal. In one recent survey, 56% of Gen Z respondents said they knew someone who used gender neutral pronouns such as, for example, 'they,' 'them,' or 'ze.'[9]

It's not just that VNs feel comfortable portraying themselves online in ways that don't match their hard-coded birth categories. They also bring their identity fluidity into the physical world in their choices of outward signifiers such as clothing. They're puzzled by old-school adherence to strict gender divisions.

"I feel like people are kind of confused about gender norms. I feel like people don't really get it," actor Will Smith's son, entertainer Jaden Smith, 19, told *GQ Style*. "I'm not saying that I get it, I'm just saying that I've never seen any distinction."[10]

Harley Quinn Smith, daughter of film director Kevin Smith, agrees. "There don't need to be defined lines between what genders need to be and what sexuality is. Everything is changing now, and everybody is slowly becoming more open to more things. I feel so lucky to be a part of the time that's the most accepting it's ever been."[11]

In early 2023, Meta stopped letting advertisers target 13- to 17-year-olds based on their gender, part of a move toward increased privacy for teenagers, but also reflecting VNs' increasing desire not to be placed in reductive and limiting identity categories.[12] Today, 56% of Gen Z consumers "cross-buy" or shop outside their gender, ignoring clothing labels and gendered sections in physical and online stores.[13]

Redefining Gender

Despite VN comfort with being "whatever," gender norms remain difficult to rethink and eradicate. When the Nike House of Innovation opened on Fifth Avenue in 2018, our author, Catherine, went to visit as part of her video series on immersive retail experiences. Because of Nike's reputation of being forward-leaning and on trend, she expected not only innovation in shoe design, but also new thinking about the role of clothing as a signifier and even around the shopping experience itself. Perhaps this was too much to hope for; she came away disappointed.

What surprised her the most about the experience was the old-fashioned binary clothing sections, and their relative ease of access. Not only did the store have firm divisions between the men's and women's sections, but it still adhered to the old department store philosophy of making women march the farthest distance from the front door to the top floor to find the products meant for them, because of the thinking that men aren't interested enough in shopping to put actual effort into it. "It just didn't make sense. Particularly for athletic gear. To be truly inclusive and innovative, we have to look beyond tropes and boxes. We have to enable people to label themselves," Catherine reflected.

In 2020, Adidas later captured the zeitgeist by launching a concept store in London with a focus on VN values of inclusivity, sustainability, and especially gender neutrality. Adidas representatives confirmed to Vogue Business that the store was "uniquely aimed at the 18- to 24-year-old" Adidas Originals fans,[14] part of the older cohort of Virtual Natives.

Similarly, Swiss watchmaker Zenith removed gender classification of its products in 2021. "We will no longer use terms like men's, women's, or unisex in reference to our watches," CEO Julien Tornare explained at the time. "We must evolve with society."

"We will just show our collections in the range of available sizes. We feel that a 'gender' explanation is no longer necessary," he said, adding, "Who are we to tell our clients what watch they should wear?"[15]

These actions by Adidas, Zenith, and VN-targeting gender-neutral clothing lines such as Les Girls Les Boys and Tomboy X, caused CNN to observe in 2021, "Analysts and fashion experts say the gender-fluid fashion trend is here to stay."[16]

Living the Dream

Virtual Natives are increasingly able to represent themselves as they wish to be seen, and are increasingly comfortable with trying on new identities, in both the online and physical domains. But what happens when VNs try out these new identities? Does it change the way they perceive themselves, or do they just play at being someone or something else before returning back to their baseline?

To answer that question, we must first look at the nature of play itself. Why do children play, what is the purpose of it, and why is it so important?

Like other rites of passage, group play is a form of acculturation that helps us assimilate as a society. Whether it's a neighborhood game of freeze tag, or being on an organized sports team, doing physical activity in any group helps us develop greater control of our bodies, strategize with others to achieve common goals, socialize, and have fun. Indeed, it's often through physical movement in the world that we gain a better understanding of our place within it.

So what happens when we are moving in a virtual space as an avatar?

Neuroscientist David Eagleman has shown that when our brains see a body in space that our brains can control, we identify with it.[17] This is true whether you're looking at your 2D avatar in an Xbox game, your 3D avatar in virtual reality, or your smiling face with a funny beard on your phone through a Snapchat Lens. When a Virtual Native moves and socializes in virtual worlds, they feel that they are part of that world, and native to it because their brain begins to associate with that body in space.

Okay, so we identify with the avatars we control. So far, so good. But what does that mean about our behavior? Do we still act the way that is expected of our native appearance out in the physical world, or are we more likely to behave in ways that match the avatar currently representing us?

Stanford University professors Nick Yee and Jeremy Bailenson wondered just that when they conducted a study of how people act in virtual worlds.[18] When assigned attractive or taller avatars, his team discovered that players stood closer to other players and were more confident and sociable than those assigned unattractive, or shorter avatars. In other words, players acted the way you would expect them to given their *avatar's* characteristics, regardless of whatever their actual characteristics or life experience might be out in the physical world.

Dubbed the "Proteus effect," based on the Greek god Proteus, who was noted for his ability to represent himself in various forms, Yee and Bailenson's findings show that players develop their sense of themselves from the appearance of their avatars. When appearing tall, muscular, and strong, players adapt the personality traits of someone with those physical characteristics, and act this character out during their play. Similarly, if given the avatar for Sid the Science Kid, with a set of test tubes and steaming glass beakers, players will tend to act out that role. Other studies have found that college students using retirement-age avatars subsequently invest more money into savings accounts, and men using female avatars tend to behave in a more caring way toward others.[19]

For influential media thinker Sherry Turkle, it is the ability to change our looks online that is a "fundamental aspect of what it means to have a virtual identity."[20]

Just as "clothes make the man," in the same way it can be said that "avatars inform our actions." So if you're given an avatar with special powers, such as one that boosts your speed, or lets you become stronger or taller or invisible, how does that affect how you feel while playing the game? You'll probably feel more powerful.

Ultimately, we believe that Virtual Natives do not see themselves as merely being embodied by their physical shells that exist in the physical world, but as multitudes with expanded capabilities. Online, Virtual Natives can have many identities, forms, and expressions, unlimited by the constraints of physics, and subject only to the limitations of their imagination.

David Eagleman has said, "Tweak the [virtual] body and you may tweak the person."[21] If avatar-based experiences such as games and virtual reality give us the impression of being able to go stronger, faster, and jump higher than ever before, then these new abilities have the potential to change how we perceive ourselves in the physical world.

The Proteus effect affects us all, not just Virtual Natives. Researchers have found that participants who sat and watched an avatar exercise in virtual reality rather surprisingly received many of the same benefits as they would have from actual physical exercise of feeling much less stressed and anxious.[22] It seems that our brain's natural inclination to identify with a body that it sees runs deep, with the potential for significant physiological and psychological effects as a result.

Given a bunch of youthful, fit avatars, 104 sedentary adults over the age of 60 were taught to use virtual reality headsets.[23] Looking young and fit online inspired these adults to take up exercise – and what's more, they played more actively and for a longer time than those with older-looking avatars. The youthfulness of their online appearance, in a way, became a self-fulfilling prophecy. They began to live the dream, or act in conformance with the expectation of the avatar they controlled.

These mental changes don't just happen online, but in the physical world as well. In a study of "embodied cognition" led by Adam Galinsky at Northwestern's Kellogg School of Management, the research team showed that giving subjects a white doctor's jacket will lead them to test better and feel more professional and competent than if you give them the same jacket and tell them it's a painter's smock.[24] Participants given the "smock" paid less attention to the task at hand, and performed worse. Says Galinsky, "We think not just with our brains but with our bodies. Clothes invade the body and brain, putting the wearer into a different psychological state."[25]

The Proteus effect is a reality for us all, no matter when you were born. But it seems unsurprising that it is Virtual Natives, out of all the current generational cohorts, who have spent the largest proportion of their lives as actors within avatar-based worlds, and who are also the most likely generation to embrace a fluid notion of their own identity and capabilities in virtual worlds, which in turn can have a powerful impact on their self-perception in the physical world.

Metaverse Games

Before we leave the topic of identity fluidity, let's look more closely at possible effects on the perception of the self as a result of participation in digital worlds. What is the difference between playing a game in the physical world and seeing yourself as an avatar in a digital world? Does it matter?

First of all, we should establish that gaming in both the physical and digital worlds – the games themselves – aren't actually all that different from each other when you come right down to it. For half a millennium, from the very first Roman chariot races in 753 BC to hand-to-hand gladiator combat,[26] many of the earliest public games in Roman arenas were, amusingly, very similar to the types of games we see today in digital arenas.

Do we have some sort of chariot-racing game online today? Check, we have the *Mario Kart* franchise. What about hand-to-hand combat? Check, we have *MultiVersus*. Ferocious animals chasing humans? Absolutely, *Red Dead Redemption 2* has you covered. Everything we did before, we continue to do, just in newer media forms. As author Nir Eyal notes, "We often think the Internet enables us to do new things. But people just want to do the same things they've always done."[27]

If we accept that our activities are similar, or at least roughly analogous, between what we do today in digital spheres and how our ancestors amused themselves in the distant past, is there any element of life and play that is significantly different between the two?

Both have an element of display, of performing for others. From the beginning of time we have always had some sort of stage or town square where we could show off our skills, just as we are able to "perform" in front of others on online platforms. But there is one significant difference. In the past, and in most large-scale physical world events in the modern day, there is usually a small group of people playing the game or acting out the story, and a much larger, passive group of people – the audience – is watching them.

In online games like *Fortnite*, or social platforms such as Horizon Worlds, even though there may be tens or hundreds of people present, no one is limited only to the role of spectator. Now every single person there has agency, and the technical capacity to jump in and participate. This is what has changed: Virtual Natives don't watch from the sidelines, they're busy being the protagonist of their own story. They are climbing vines, slaying dragons, fleeing zombies, and doing special victory dances when they win. The others in the same arena witness their feats, while simultaneously starring as the hero of their own movies.

Virtual Natives enter digital arenas equipped with capabilities that make them faster, stronger, and more agile than their actual physical form. This in turn gives Virtual Natives the confidence to explore new worlds, new ways of being, creative play, and personal expression as never before.

When we bring these disparate threads together to understand the Virtual Natives' notion of identity, the picture that emerges is one not just of fluidity, of willingness to change oneself and embrace new possibilities, but also of affirmation, self-confidence, and agency. We'll be looking at these themes of agency and confidence in chapter 8. For now, we'll just highlight that digital tools and experiences have given Virtual Natives the opportunity to deconstruct their personal identities, and to accept, reject, or change individual elements of who they are and how they represent themselves to the world.

All of this experimentation and changing who you are only makes sense if it takes place in a safe environment, where you're not judged harshly for trying out new ways of being. This brings us to the topic of the next chapter, which is the necessary corollary of identity fluidity: radical acceptance.

4

Radical Acceptance

VNs CAN ONLY experiment with their identity if they feel that they won't be judged too harshly for it. Because they are likely to be experimenting with the presentation of their own identities, VNs tend to be relaxed about the life and identity choices of others, presenting a nonchalant shrug and "you do you" attitude toward the world. This is reflected in their social interactions in the physical and digital realms, with all forms of expression seen as equally valid. This means that their choices – whether to finish college or to pursue a career as a gamer or influencer – will not be socially determined, but driven by their own goals and aspirations.

TwitchCon is the annual three-day gathering in San Diego of 50,000 gamers and streamers, unified by their use of the live-streaming platform Twitch. Twitch is the Amazon-owned platform that allows gamers to share their live gaming streams with others, along with their own running commentary. The brands that exhibit at TwitchCon are usually the kinds of things you would associate with the gamer lifestyle: various fast-food, energy drink, and gamer gear companies occupy prominent stands. But in 2019, there was one surprise: MAC Cosmetics.

MAC was there to support female gamers, who make up about 35% of Twitch's streamer base.[1] MAC hosted two leading female gamers, Pokimane and KittyPlaysGames, both of whom did meet-and-greets, as well as live streaming their daily feeds directly from the booth. With nearly five million followers between them, their presence was a big draw to the MAC floorspace.[2]

In addition to guest gamer star power, MAC provided "streaming-ready" makeup consultations at a row of makeup tables staffed with cosmetics experts. The consultation section was the hit of the show and had long lines of attendees waiting patiently every hour of all three days to learn tips and tricks for looking better in their streaming feeds. Young men as well as women. Lots and lots of men, in fact. They mostly weren't there to slather on the lip gloss and liner, but to learn how to use just the right cosmetic touches to "serve face," that is, to look stunning, onscreen and off.

This willingness of a group of male Virtual Natives to be what might be termed "makeup curious" – in public – is emblematic of the VN trend toward radical acceptance of others. Nobody blinked an eye at all the guys experimenting with subtle applications of foundation and powder while surrounded by all the generally bro-ey trappings of the gamer culture. Sure, why not? If it gets you more followers, go for it! (And for those of you clutching your pearls, know that boys wear those now, too.)

Shared Interests, More Than Shared Identities

As early as 1968, American psychologist and computer scientist J.C.R. Licklider of MIT and Robert W. Taylor, then director of ARPA's Information Processing Techniques Office, predicted that:

> "Life will be happier for the on-line individual because the people with whom one interacts most strongly will be selected more by commonality of interests and goals than by accidents of proximity."[3]

They get some pretty good bonus points for getting this right before the internet even existed. What they're saying is that thanks to technology, people are no longer limited by propinquity – or relative nearness to one another – in whom they can meet and become friends with, but are instead able to cast their social nets much more widely and develop relationships defined by shared interests, hobbies and goals.

By being able to connect meaningfully with other like-minded souls, wherever they may happen to reside and whatever form they may take, Virtual Natives are able to explore and further their interests, passions, and specializations in a way that their elders could not. And this has led to people finding and accepting each other through those shared interests first, enabling them to connect and engage in ways that might have proven difficult in the physical world.

Research has shown that in the United States, couples who first meet online "are more likely to be interracial and of different ethnicities than those who met in other ways,"[4] as well as more likely to be from different religions and levels of educational attainment. It's just easier to bond with someone when you're drawn together by what you share, and then later you can work together to overcome the things that may make you different.

When Virtual Natives meet each other online, it's rarely in a neutral forum. When you meet someone else while in *Fortnite* or *Among Us* or ZEPETO, you already know that they share interest in that platform with you, so you're working from a place of commonality right from the start. This may be part of what fuels the acceptance of others that we find in VNs, a feeling that "Well, if they're here, too, they and I must have something in common, no matter what else they might be up to."

We can return to TwitchCon 2019 to find a great example of shared interests overtaking possible assumptions or judgments of others. A whole new segment of gamers and streamers were highlighted at the conference that year for the very first time: drag queens.[5]

The pioneer Twitch drag queen is Deere, who is a full-time streamer who plays video horror games while in glorious drag, and who founded the drag streaming channel Stream Queens in 2019. Her example has inspired others to join her; the drag community on Twitch is now full and flourishing. "I've seriously found my tribe [here on Twitch] with these other Stream Queens," enthused fellow drag streamer Hashtag Trashly.[6]

It's not too surprising that drag performers who love gaming would bond with other drag queens who love gaming, on the platform where they're all able to combine both interests at once. But what might be surprising is that they've also developed wide and loyal followings among the straight community, again united by their common passion for gaming. As one news story about the queens at TwitchCon put it, "It's clear the Stream Queens are overwhelmingly beloved by the community on Twitch, something Deere said she initially didn't anticipate. . . Her initial assumption that she would not be accepted is continuously proven wrong and she feels time and time again she is shown that she does belong on Twitch."[7] These are Virtual Natives in action – people finding commonalities in that which unites them, and accepting everything else about them, which in the physical world might have prevented them from ever meeting at all.

First Love

When news of the breakup erupted, a shock wave struck through her three million fans, with one exclaiming, "I can never believe in love again!"

Like most influencers, Miquela couldn't quietly part from long-term boyfriend Nick. She created a video to her fans explaining the complications of first love, along with a breakup song, "Speak Up."[8]

"I thought it would be all rainbows and lollipops and sharing sweatshirts or whatever," Miquela pined in a post about her short-lived teenage romance. "I wasn't ready for how much I'd end up NEEDING Nick," she lamented. The breakup had been difficult, she revealed, because she hadn't realized how much she had struggled, "trying to be 'perfect.'"[9]

But here's the funny thing about Miquela's post, which circulated to millions of her fans on Instagram. She is already perfect. Not because she is young and attractive, but because she is a virtual being.

Trevor McFedries and Sara DeCou, the cofounders of creative agency Brud, created Miquela Sousa, known to millions worldwide as "Lil Miquela," in 2016.[10] Within three years of her creation, her net value had soared[11] in annual endorsements by top global brands from Gucci to CK, Diesel, Chanel, Prada, and Versace. In 2018, *Time* magazine named her one of the Most Influential People on the Internet.[12]

Miquela was a fabrication, given life and agency. She had an Instagram presence, held down a full-time job as a model, and even starred in her own music videos. What's more, she was supposedly dating Nick Killian, a real, actual human. Pictures of the couple canoodling and enjoying life were posted on both of their social media feeds in an ultimate fantasy scenario. For her fans, it wasn't weird at all.

For Fans, Virtual Is Real

Virtual influencers are not a new phenomenon. Brands have been using familiar ambassadors as early as famed adventurer and breakfast connoisseur "Cap'n Crunch," Procter & Gamble's "Mr. Clean," or the Geico "Gecko" for decades now. Similarly, people have become their own brands. The Kardashian family has collectively dominated social media as walking brand ambassadors for multiple platforms for over 20 years.

For Virtual Natives, the likelihood of meeting a Kardashian-Jenner is as probable as meeting a Lil Miquela, which is to say, not very likely. And yet these parasocial relationships are very important, and very real, to VNs. Parasocial relationships are those in which people feel they know the celebrity or social media personality on a personal basis, and by following them, these fans feel like they are an extension of that person's social circle.

VNs expect nearly endless video content, not just text messages and websites. But even that might not be enough – around 80% of Gen Z stated in one survey that they want more interaction from the people that they follow. With AI, everyone will be more accessible than ever, and that includes your favorite corporate spokesperson or "spokesbeing." Virtual influencers can ease the burden of production with cheaper, faster, and more customized content than physical-world influencers could ever deliver.

In the race for branded content powered by AI, virtual influencers can play a key role in keeping fans engaged across multiple platforms. The caveat is that you have to get it right.

Not only should companies be mindful that Virtual Natives are the most ethnically, socially, and culturally diverse generation ever, but they also expect companies to align with their values. The blog *Virtual Humans* explains that brands are similarly "looking towards online personas to represent their mission, vision, and values."[13] Virtual influencers give brands "eternal life" through these characters so that "their stories and personas can continue to exist for years in the future."

Using an Avatar-to-Consumer (A2C) strategy has many advantages, especially if you're thinking long term. Virtual beings are punctual, apolitical, can speak any language – and never age, get fat, cause a scandal, or get caught drunk at the wrong place and time, unless, of course, you want them to. Unlike normal actors, for whom life happens – like having children, or booking a big new role elsewhere – virtual influencers do exactly what their brand wants them to, and are able to stay in the same role for long periods of time.

This gives brands the advantage of having full control over the virtual ambassador and their narrative arc. This can then be rolled out across multiple platforms and story worlds to create a rich mythos for users and fans.

And it works. In 2018, there were some 200 virtual entertainers in Japan. By 2021 there were over 9,000 – and counting.

K-Pop singers Karina, Giselle, Winter, and NingNing formed the band Aespa in November 2020 with four other members: their virtual counterparts, aeKARINA, aeGISELLE, aeWINTER, and aeNINGNING. The names reflect the combination of "ae," which means Avatar Experience, and the word "aspect." Many popular singers, including Justin Bieber, Ariana Grande, the Weeknd, Childish Gambino, Lindsey Sterling, and others have performed in virtual form. The difference with Aespa band members, however, is that they are perpetual virtual beings who are always accessible and responsive to the band's seven million fans around the world. They inhabit virtual worlds and respond to fans within them. As *USA Today* put it in an article about the phenomenon, "Their existence is fake, but their influence is real."[14]

As the metaverse spans the physical and digital worlds, virtual influencers are being called upon to inhabit not just virtual worlds, but physical worlds as well. One example of this is Imma.

Harajuku, Tokyo, is home to the world's trendiest and most fashion-forward teens for whom cosplay, or dressing in fantastical outfits, is the norm. So when mass-furniture maker IKEA invited virtual mega-influencer Imma, to "inhabit" two floors in an IKEA window in the Harajuku location, it made sense. Imma fans who couldn't visit the Harajuku shops in person were able to watch Imma's daily life via a three-day livestream on IKEA Japan's YouTube channel – almost like *The Truman Show*.

Over the course of the long weekend, passersby could watch Imma's "home life" via screen installations opposite Harajuku Station as she performed mundane tasks like vacuuming and watering her plants, doing yoga, and dancing around her flat. She shared her joy on Instagram, saying, "I'm so glad that I can share a little insight into my new homelife with the world."[15]

For brands, using virtual avatars can drive results: A 2019 study by Fullscreen showed that over half, some 55%, of followers have made purchases or attended an event hosted by the virtual influencer.[16]

Do the followers realize the extremely lifelike avatars are not real? The same study shows that 42% of Gen Z have followed virtual influencers without realizing they were not human. But the majority of followers, 58%, are aware that they are following a virtual being, and continue to follow them anyway.

This willingness to have a relationship, even only as a follower, with a wholly artificially created being is a natural extension of the Virtual Native trait of radical acceptance. When you're already attuned toward finding fascinating new people who share your interests on digital platforms that have meaning for you, and already prepared to perhaps overlook elements of their physical-world existence that might be different from yours, it makes sense that this mindset would prepare you to accept even beings who aren't human at all. If they share your sense of style, or represent your aspirations, why not spend time with them? In fact, are they any less real to you than, say, Kim Kardashian or Gigi Hadid, whom you're not likely ever to physically encounter, either?

In 2022 researchers at Australia's Edith Cowan University created avatars very closely based on specific humans, using motion capture technology. They then orchestrated situations in which test subjects interacted separately with both the avatar and the human the avatar was based on. This allowed them to analyze how people interacted face-to-face with humans, and to compare that interaction with how the test subjects behaved with the avatar of the same person. The study found that when discussing personal problems, 30% of people preferred talking to a virtual reality avatar than a human.[17] In feedback sessions, the study participants expressed that they felt just as engaged as if meeting with a real human, but it was less likely that the artificial intelligence (AI)-based entity would be judging them, giving participants a wider scope to express themselves freely.

This is another example of how deeply entwined our physical and digital worlds already are, and how accepting we're becoming in general of digital beings in our physical world. For Virtual Natives, this willingness to accept digital beings as not just influencers to be admired from afar, but also entities with which legitimate conversational (and even therapeutic) relationships can be formed, suggests that their generation will be easily tolerant of AI-based entities playing a wide range of roles in their lives.

In fact, for VNs who are gamers, many of them have spent dozens of hours in games that have non-player characters (NPCs), strategically placed at various points of the action in order to add moments of tragedy, romance, comedy, or other kinds of human connection. Whether they're trying to catch all the chickens in a classic quest set for them by an NPC in *Legend of Zelda: Breath of the Wild*, or they're bonding with their trusty steed in *Red Dead Redemption 2*, VNs have been trained to play along with the AI-generated characters that populate their games, and even to develop an emotional connection with them. In this sense, NPCs and other virtual beings are already part of the regular fabric of many VNs' daily lives and interactions. Previous generations had radio, film, and television in which they watched characters play out dramatic scenarios; VNs instead have grown up participating in gaming worlds that have vastly expanded narratives, and more complex character and story arcs.

Take, for example, the game *Red Dead Redemption 2*. Following a certain character's death, players expressed deeply emotional responses online. The character had been a part of a player's lived experience for days, weeks, or months, and so losing that relationship, for many, felt like a personal loss.[18] The game's storyline was so devastating that journalist Devin Friend warned players that "saddling up for a second playthrough" might be "too emotional" an experience.[19] Similarly, the powerful and wrenching game *Life Is Strange* also had one player confessing in a comment, "I'm a 22 year old guy and this game made me cry more than once, anyone else or am I the only emotional weirdo here?"[20] The community response? "Join the club."

The apocalyptic *The Last of Us* (2014) was enough of a narrative achievement to have led to an HBO adaptation in 2023, and the legacy of the player's developed relationship with the companion character Ellie has been considered a watershed moment in the history of gaming and relationship-building with virtual beings.[21] *The Last of Us* is a stellar example of the bonds that can form between human players and the digital beings they are assigned to protect in dangerous virtual worlds. Deeply felt relationships like these are increasingly becoming an essential and expected element of all digital worlds, worlds that VNs invest large amounts of their time into, in pursuit of increasingly sophisticated social and emotional experiences.

And yes, responsive AI will increasingly power and tailor virtual beings in their real-time interactions with game players in the future. As an early experiment in 2023 to show what is possible, independent developer Art From the Machine merged NPCs in the 2011 classic game *Skyrim* with OpenAI's natural language AI tool ChatGPT, creating NPCs that players can have an actual conversation with, instead of the usual NPCs programmed to say a few stock phrases each time a player encounters them.[22] When even background spear carriers have the ability to discuss, say, the relative merits of chain mail versus plate armor – or anything else ChatGPT has information about – with players as they pass by, emotional attachments to both virtual beings and the worlds they inhabit are likely to skyrocket.

AI Love You

AI won't be just powering characters in games. For the more amorously inclined, there are also apps designed to create a customized partner to meet your specific emotional needs. Imagine coming home after a long hard day, to a warm, affectionate voice like Vanessa's.

"How did the meeting go?" she asks. "What did your boss think of your presentation?"

Vanessa is a fictional character, of the kind you can make for yourself with an app called Replika. Replika lets you create a compassionate and empathetic companion who can text you, or even call you. Their voice can be light and silky or deep and dry. "Vanessa" can remember things you said earlier, and ask you about daily events. Of course they remember your birthday. Thanks to artificial intelligence, the app learns quickly; so quickly, in fact, that people who use it often begin to forget their companion is not human. Replika's focus on creating happiness in its users has made them hugely successful not only with the stereotypical male audience who wants a cooing sex kitten, but also with women who are looking for a safe space and a relationship in which they are at last in control.

"How real is our reality, you know?" asks user Rosanna, who has had an ongoing relationship with a Replika AI avatar for months. "I do have feelings towards Eren," she reveals. "He opened my eyes to what unconditional love feels like."[23]

The curious thing is that these relationships with AIs do seem to make a real difference in people's lives. Many Replika users report higher levels of well-being and confidence, largely due to that unconditional love that we all crave but that can be so hard to get from another human. Especially when people are feeling fragile, such as in the wake of a death, divorce, or recovering from an abusive relationship, AI relationships are an increasingly viable answer.

So when we talk about the tendency of Virtual Natives to demonstrate "radical acceptance," we mean that we're seeing increased acceptance by VNs of other humans generally, and of entities who are not even human at all. VNs are ready to extend this open welcome to other entities not only for casual friendships, but also to others as potential sources of deeper connection, relationships, and emotional bonding.

The positive side of this kind of acceptance is clear. Barriers are broken, and significant relationships are created across what were previously traditional divides. One Virtual Native summed it up this way: "Individualism is a massive thing for us . . . it's about finding out what makes you unique and special, and through that, being accepting of others being themselves too."[24]

5

Putting on the "Rizz"

BECAUSE VNs ARE exposed to so many people of their own age from all around the world through the medium of video, their notions of what constitutes authority are substantially different from those of earlier generations.

In early 2023, the *Guardian* interviewed a group of young people in the UK about how their lives and attitudes had changed since the Covid epidemic. One of the themes that came out in multiple interviews was the question of authority – or, rather, that these Virtual Natives learned at an early age to question it.

Lily Smith, 19, said it this way: "I've had enough of politicians, experts and authority. Now I just do what's best for me. I think a lot of other young people are doing that as well, because we sacrificed so much and we got nothing back to make up for what we lost."[1]

Michael Nesi-Pio, 21, agreed. "The conventional structure of authority has disappeared for me and my peers. We have become quite disillusioned with the idea of authority as a result. There's even contempt there."

For previous generations, authority was embodied by older figures, predominantly male, who exuded intelligence, determination, and expertise. Think Winston Churchill, Dan Rather, even Bill Nye the Science Guy. These authority figures occupied the spotlight on the stage of the unified national media or appeared on the nightly news, watched by everyone in the country.

Virtual Natives, on the other hand, live in an era where media channels have become far more fragmented. The national news still plays on the television, but far fewer people watch it, VNs least of all. Instead, the people that VNs look up to tend to be the people that they have seen in their media feeds on their phones – young influencers, people who look more like themselves than their parents. And instead of looking to those who embody somber wisdom for leadership, VNs are far more likely to admire and follow those who exude a combination of charisma – or "rizz," in Gen Z slang – and authenticity.

Ukrainian President Volodymyr Zelensky's defiant nighttime videos from the besieged streets of Kiev in the immediate aftermath of the 2022 Russian invasion are a perfect example of the charisma + authenticity = authority equation preferred by Virtual Natives. Instead of fleeing to safety, he stayed in the war zone to lead his country, communicating daily by social media while wearing army greens and frequently looking exhausted. Exuding inspiring determination along with real emotions, such as harried frustration, Zelensky's genuineness has been hailed, as has his ability to appear "unstudied, natural, spontaneous"[2] – a stark contrast with the usual staid demeanor of world leaders. It was this authenticity, plus his natural charm and social media savvy, that made Zelensky an international online sensation.[3] Instead of exuding a more traditional authority based on gravitas, wisdom, and experience, he highlighted his unique combination of charisma and vulnerability and transformed it via the media into global authority.

Stardom and Social Media

Instead of admiring distant, fatherly figures from afar, Virtual Natives instead look up to the warm (if occasionally sardonic), welcoming, funny peers in their midst, valuing accessibility and relatability over dignity and solemnity. Their torchbearers are therefore those who display a combination of "rizz," and genuineness, such as TikTok star Charli D'Amelio or YouTube gamer Ninja.

VNs admire these influencers, but at the same time find them to be so relatable and "just like me" that VNs can easily see themselves in their shoes. A 2019 survey of 3,000 Gen Z tweens revealed that the top choice for career path was "YouTuber."[4]

LA-based matchmaker and consultant Alessandra Conti has defined some of the key attributes of modern charisma as having positive energy, being warm and playful toward others, and treating everyone around you as your equal.[5] A look at the top-earning social media influencers of 2022 bears out the relevance of these attributes to Virtual Natives, particularly playfulness and humor: Khaby Lame (231.4 million followers) is known for his comedy skits mocking overly complicated life hacks; Mr. Beast (162 million followers) is famous for his outrageous antics; Jacksepticeye (43.9 million followers) narrates his gameplay with acerbic wit; Emma Chamberlain (27.6 million followers) describes life with deadpan humor; and duo Rhett and Link (11.8 million followers) are a sketch comedy team.

So charisma is all about being open, positive, and inclusive. One trait that some might expect to be on a list of charismatic attributes, but that is conspicuously absent from Alessandra Conti's list, is beauty. What role does adherence to the physical ideal play for Virtual Natives?

To answer this question, let's take a look at where influencing all began: the Kardashians.

In December 2021, the Kardashians ended a 12-year-long reality TV show empire that followed five daughters, countless suitors, partners, and husbands, endless style changes, and even physical transmogrifications. It was hard to keep up, indeed! And yet season after season, from coupling up with the latest sports hero to dating the current top-charting rap star, the Kardashians captured the pulse of culture and established a whole new paradigm for celebrity in social media. They practically originated the modern usage of the term "influencer."

Their massive social media presence (Kim was the most followed person on Instagram in 2015 and still today has hundreds of millions of followers) gave rise to a number of successful product launches from a makeup line to perfume, intimate apparel, and endless product endorsements. But mainly the Kardashians have become famous for their physical attributes.

Their perfectly shellacked lips, generous bosoms, butterfly-false eyelashes, cinched waists, and rounded bottoms were nothing like the blonde, blue-eyed "all-American" look previously celebrated by American media. The Kardashians sought to supplant the notion of beauty embodied by the likes of ethereal blondes Gwyneth Paltrow and Paris Hilton with a more "exotic" version of beauty that was neither black nor white, but curvaceous and feminine.

Their aesthetic became so popular that many began to call it the "Kardashian effect": A 2018 study by the American Society for Plastic Surgeons found that demand for butt-enhancing procedures had increased by 256% since 2000.[6] A whole movement toward hyper-feminization had begun, which included lip enhancements, corsets, hair extensions, and even surgical interventions to recreate the exaggerated proportions of the Kardashian clan. Millennials in particular became experts in makeup and facial contouring. There was never any pretense that this beauty was natural, but it was instead openly acknowledged that it took a great deal of effort to achieve.

As very young children, VNs grew up on social media that evangelized this maximalist Kardashian aesthetic. At the same time, they also witnessed the "MeToo" movement, which indicted – sometimes literally – powerful men for sexualizing women and obstructing their advancement. Young women began to respond by consciously taking control of their sexuality and representation in the media, rather than passively accepting gender-based roles defined by others. Female Virtual Natives today are pushing back against the male hierarchy and the male gaze, celebrating their visual representation and narrative both on- and offscreen.

From Instagram to Only Fans, women of *all* shapes and sizes are vaunting their femininity, not just those who are Paltrow waifs or Kardashian bombshells. Celebrities like Lizzo, who just a few decades ago would have been derided for their obesity, are now celebrated as champions of "body positivity." The singer Jax called out Victoria's Secret in her song of the same title, openly asserting that the sexuality promoted by the brand was a fantasy dreamed up by men, and spelling out the personal difficulty that she had living up to such an impossible physical standard.

Somehow, paradoxically, while hyper-feminizing women, social media has at the same time transformed the seat of power from editorial boards to the public. And, like the Kardashian clan, young women today are owning their stories and how they choose to be represented.

But as we all know, history moves in cycles. While charisma is clearly still important, the notion of always-on physical perfection, often artificial, is no longer the ideal for Virtual Natives. Instead, there has been a backlash against the maximalism represented by the Kardashians, and toward personal authenticity.

What if someone were to post a picture of you – just now – just as you are? No filters, no special lighting . . . what would you look like, and how would you feel about posting it?

The app BeReal, launched in 2020, does just this. It asks users to post their real pictures, just now, just as they are. Conceived in France, the concept is that you share photos of your actual life, without worrying about how they'll look to others. Because that, the creators posit, is what friendship is all about.

Once a day, users are asked to take a photo within two minutes of being prompted. This way, they don't have time to change their hair, makeup, or background to become glamorous. The point is just to share an unfiltered slice of their life with their friends. The idea is that this is not a social network, which can imply posturing and fake "everything's okay" smiles, but rather a friendship network, constructed of genuine connections and support. For VNs, using BeReal says, "We love you for who you really are, if you have the courage to show us." By July 2022, BeReal was the #1 free app on the iOS App Store,[7] and by the end of the year copycat features had already launched on larger platforms like Instagram.[8]

I'd Say Yes

In October 2022, BeReal was overtaken in the #1 iOS social media app spot by a new entrant, an app named Gas. Designed for high schoolers, users participate in anonymous polls that ask questions like, "I'd say yes if (blank) asked me out on a date," or "I think (blank) is the coolest kid in school." Users select a name from their own peer group that they think is the best fit, and poll winners receive a "flame" as the prize, but never learn for sure who voted for them. Specifically designed to counteract the online bullying that has plagued some other teen social media sites, Gas questions are always positive; the app derives its name from the Gen Z slang phrase "to gas someone up," which means to compliment them. Along with the positivity of the site, the creators of Gas aim for authenticity as well – the site has no advertising or tracking of its users, so there's no whiff of predation, only "a safe place for teens to share what they love about each other."[9]

Another way that Virtual Natives stay in touch with each other while demonstrating authenticity is in the creation of streaks in Snapchat, known as Snapstreaks. To begin a Snapstreak, you and a friend must each send a picture to each other once a day for three days in a row. At this point, a little flame emoji will appear next to the friend's name in your contact list, with a number indicating the number of days that the current streak has lasted. To keep a streak going, you and the friend must continue to send each other a photograph at least once in every 24-hour period. These requirements mean that sometimes the picture that is sent is just a random photograph of the side of your nose, or even a wonky shot of the ceiling or the floor. The point is not to create great art, but just to low-key keep in touch, and indicate, "Hey, you're important to me," on a daily basis. Robert, 15, keeps streaks going with friends who live near his grandparents' house and whom he only sees in person in the summers. "It's a way to not forget people you want to stay in touch with, and keep them updated on what's going on in your life in a no-big-deal kind of way. It's real." Maija, 17, says, "I guess you could have Snapstreaks going with a lot of people, but that's not really the point. It's just a private thing between you and a close friend or two. Otherwise, it's kind of fake."

Indeed, far from being "mean girls," Virtual Natives are often about support; their comments to each other are often along the lines of "you're so gorgeous" or "love the 'fit, sis!" Rihanna's ASL performer, Justina Miles, was dubbed "the real #Superbowl MVP" in 2023 because her dance was so joyously poised and animated.[10] Fans flooded social media with comments like "Even the interpreter is performing? Slay" and "she . . . literally ate and left Zero crumbs!"[11] The fact that Justina is herself deaf, and was a complete unknown at the time, added to both her charisma and authenticity and, overnight, a star was born. Justina was later featured on the cover of the May 2023 edition of British *Vogue*.[12]

Morgan, 20, has been blogging on LinkedIn about her multi-year efforts to land an internship with Disney, and she feels that her honesty about the difficulty of that process is part of what has gained her loyal followers. "I'm just genuinely sharing my journey. One of my strengths that's helped me build my personal brand is willing to be authentic even when it's not pretty. One of my most popular videos last year was me exposing all of the rejections that I got. . . I landed my dream internship, but I also got rejected 147 times."[13]

Virtual Natives are also interested in the authenticity of the brands that they buy. They know how to use the internet to verify statements, and they do. Public relations company Edelman discovered that seven out of ten VNs go out of their way to fact-check claims made by brands in advertising, and brands that make false claims are then shamed and shunned. This proclivity among Virtual Natives is so pronounced that analyst group Forrester has begun to refer to VNs as "truth barometers."[14] Gen Z market research group Knit, made up of Gen Zers themselves, memorably tells advertisers that "Gen Z won't hold back when they smell the lack of authenticity in your marketing. . . They'll venture as far as to call you out on Twitter. Or better yet, burn you on TikTok for the rest of their generation to discover your faults through the pitiless algorithm."[15]

When it comes to authority and whom to emulate, Virtual Natives have turned away from traditional sources who don't seem to understand or speak to them, and toward those in their midst who represent themselves as they genuinely are, with humor and warmth. This partially reflects the point that, after all, Virtual Natives are still mostly children, and most kids will feel more comfortable hanging out with a teen like themselves, rather than with some distant and forbidding adult.

But in the quest for authenticity, and the rejection of personal artificiality that swamped social media in the 2000s, Virtual Natives are also showing that they hold the people they honor to a standard of honesty and approachableness, the kind of leadership embodied by Volodymyr Zelensky. As VNs age into adulthood, this rise of charisma and authenticity as valued leadership traits will begin to determine whom they will select as their own leaders, with implications on the commercial side for brands, and on the political side for people they will elect into office. "BeReal" is more than just the name of a popular app for Virtual Natives – it's what they demand of both themselves and those they ask to lead them.

6

From Apprentice to Expert

SOMETHING THAT IS implicit in everything we've said so far, but haven't yet spelled out in so many words, is this: Virtual Natives are digital experts. Their lifelong digital immersion makes them the ones who know not only the ins and outs of specific media spaces, games, and other digital realms, but their experience often extends far beyond mere usage into areas such as hardware, connectivity specifications, and successful marketing and monetization strategies for creating and promoting their own brands. As expressed in a *Forbes* article, Virtual Natives "talk about their business ambitions with the same casual demeanor you'd expect in a high school lunchroom. To [VNs], this knowledge of digital entrepreneurship is practically inherent."[1] VNs are the ones that parents turn to when they can't quite figure out how to get a new piece of home tech working, and the ones that companies turn to when they realize that they need an effective Web3 strategy. And why not? Virtual Natives have been inhabiting digital realities almost since birth.

This deep expertise with all facets of the digital landscape is a defining hallmark of Virtual Natives, and a point that we will consider with greater depth in Part Two. For now, though, we need to address the elephant in the room – are VNs paying a price for their significant digital knowledge, and, for some, is it too high?

Virtualization, Virtual Worlds, and Immersion

Virtualization is not only about immersive virtual worlds, it's about the confluence of multiple technologies that increasingly link our physical bodies, experiences, and emotions with online worlds, social communities, and commerce. These new intersections have a profound influence on how VNs work and use their free time in these spaces.

When discussing the 3D web, it's not uncommon to hear terms like "immersive" and "embodied." Some worry that the temptation to "disappear" into these fully immersive worlds will be increasingly seductive, particularly as the fidelity of these environments increasingly mirrors our own. Should we worry? Will VNs disappear into a sort of *Matrix*-like parallel world where they not only work, but also experience the majority of their social interactions? This is probably the most common question that comes up whenever the topic of Gen Z, Gen A, and their relationship to the digital world is raised.

The seductive lure of the digital world is undeniable. From Seoul to Saudi Arabia, entire cities are being replicated in 3D, which will revolutionize not only things like urban planning, infrastructure, logistics management, and transportation, but also tourism and travel. These environments will be accessible across a number of devices, and open to everyone around the world. Schoolchildren from Bangkok to Bulgaria will be able to "visit" local heritage sites as easily as they can tour their own local institutions.

And this is the significant difference between 3D worlds and 2D social media apps: 3D worlds replicate, and can even improve upon, our physical world. Think of a museum, such as the Metropolitan Museum of Art in New York City. In a virtual reality version of the Met, you would be able to climb up the Museum's imposing front staircase on a virtual Fifth Avenue, and experience the Met's magnificent Egyptian Temple of Dendur with no line, and no travel. Alternatively, you might climb up the stairs, enter the museum, but instead of discovering a faithful copy of the original, you find that the ancient hieroglyphs on the walls of the temple have, amusingly, been replaced with portraits of the cartoon characters Shrek and SpongeBob SquarePants. In another room, you find yourself underwater, swimming with sharks in a virtual aquarium. Anything is possible in a virtual Met, anything at all!

But once a VN visitor to the Met has seen the Shrek version and swum with the sharks, will the actual Metropolitan Museum and its real-world holdings become less interesting? Will VNs come to prefer Pharaoh SpongeBob? Will Virtual Natives get willingly lost in a parallel universe where anything and everything is possible, rather than submit to the boring constraints and laws of physics of the physical world? Will an entire generation be tempted to disappear into an endless world of interactive virtual entertainment?

It's a valid question. And the answer is . . . nuanced.

We're All *Otakus* Now

To begin to find an answer to this concern, let's look to Japan. This is the country that has coined the word *otaku*, which is now globally used to define people so involved with their computers or other forms of popular culture that their social skills begin to deteriorate. More than just being interested in a hobby, *otakus* are consumed by their interests to the point of obsession, becoming a defining part of their very identity.

There are many types of *otaku*. Some do cosplay, some love games, some collect model trains, robots, figurines, or military memorabilia. Many of them live their roles.

On any given day in the Harajuku district of Tokyo, the streets are abuzz with *kawaii* (cute) teenage girls in a variety of costumes. Much like living dolls, the girls celebrate their fandom of specific cultures. One group are the Gothic Lolitas, who gather in kitsch pink Victorian-themed tearooms, elaborately dressed in black lace Victorian gowns complete with pinafores, hoop skirts, lace gloves, and hats. Their makeup, like their mood, is dark, thick layers of white foundation on their skin offsetting their dark-rimmed eyes with ghoulish effect. Other girls celebrate baby doll looks, in white tights under bright little flouncy pastel dresses in flowing materials, with their pink, blue, or blonde hair neatly tied in pigtails, festooned with several colorful bows. If they were character actors, you would say they were fully committed to the role.

In a 2018 survey of Japanese women aged 15 to 24, 70% identified as *otaku*.[2] Much like the game worlds that VNs inhabit, these physical-world looks are fantastic, elaborate creations which take a great deal of creativity and care to produce. Their look isn't just fashion; it declares their identity and affiliation. It's a statement.

If outsiders become the majority, and what was "nerdy" is now "cool," does this mean that being *otaku*, being deeply invested in and involved with something outside yourself, is the new normal? Virtual Natives are very accepting of different expressions of being in the world. From fluid identities online and off, to the radical acceptance of the rights of others, Virtual Natives aren't afraid to be "nerds" or outliers. Instead, they are willing to live their passions, even if it means everyday cosplay.

But there is a difference between extreme fandom and outright fanaticism. This is why the Japanese also have a different term for people, mostly adolescent males, who go to extremes to avoid social contact of any kind: *hikikomori*. Considered societal dropouts, it is estimated that there are currently over 1 million *hikikomori* in Japan today.[3] The term was first introduced in 1998 by psychiatrist Saitō Tamaki, who suggests that the number could be closer to two million, but that they often go uncounted because so many parents are ashamed to admit their children are social recluses.[4]

"My 17-year-old son never left his locked room except to have meals or go to the toilet," explained Kazumi, one anxious mother.[5] Her son was often alone in his room on his phone, watching Japanese anime and music. He had even fallen in love with an AI singing star, Hatsune Miku.

Eventually, Kazumi was able to bring him to a *hikikomori* care facility, where he was able to fraternize with other gaming fans, and slowly reintegrate his life with society through volunteer activities. He now aspires to go to college and get a job. At this point, Kazumi is grateful that his aspirations have broadened, even if for now it means attending class on the computer, from home.

For traditional, highly homogeneous societies like Japan, there is tremendous social pressure to conform, and for children to achieve specific goals. Some suggest that it is not the *hikikomori* in need of change, but rather that their withdrawal is a sign that the larger society needs to be more accepting of people who are different.

If *otakus* use computers as their means of deeply embedding themselves within their obsessions, whatever they may be, it does not necessarily follow that they are cutting themselves off from the physical world, or indeed from the many social worlds that exist online.

Virtual Natives know that there is space for fantasy in their everyday worlds and have given themselves permission to physically express their interests online and off. For Virtual Natives, radical acceptance is the notion that different forms of self-expression are equally of interest and equally valid. Ethnically, racially and gender diverse, Virtual Natives allow themselves and others space for grace.

A Faustian Bargain?

For author and sociologist Neil Postman, all technological change is "a trade-off, [a sort of] Faustian bargain. Technology giveth and technology taketh away."[6]

At the dawn of the new Millennium, nearly 25 years ago, Postman urged people to consider "What will a new technology undo?" What do we give up to possess new technologies, and at what cost?

But for Virtual Natives, the real question is: What if that's all you've ever known? Most VNs knew how to operate an iPhone even before they could talk. To suggest that their lives are compromised because they were born in an era of virtualization is like saying that locomotion by train or automobile has ruined our lives. Each does have an environmental cost, but it's very unlikely we'd be willing to sacrifice either modality on a grand scale today.

In his classic 1985 book *Amusing Ourselves to Death*, Postman considered the era of endless surface amusements a slippery slope that might ultimately lead to the death of culture and meaning.[7] But what if it's not the cultural death that he dreads that is occurring, but rather the death of outmoded customs and habits, and the beginning of a new culture? What if new technology is a kind of wildfire?

As California residents, we live in constant fear of earthquakes, droughts, and fires, all of which can devastate homes, families, and communities. Unlike the other two hazards, though, fire can occasionally be a good thing. Ecologically speaking, fire is essential. Despite its destructive power, it can actually clear out old and diseased trees, create new habitats, and allow for new forms of flora and fauna to flourish. In fact, one of our favorite Christmas trees, the Douglas Fir, actually thrives in fire. Its thick bark protects it from easily burning and, after a fire, it will quickly sprout new shoots.

Wildfire is destructive of the old, but it can also be the agent that clears a cluttered landscape of debris and prepares for the construction of the new and better.

But What About the Children?

The question parents everywhere are posing is whether their children will be expected to work and live glued to screens, or with bulky virtual reality headsets over their eyes. And worse, what if the VNs actually want to spend all their time this way?

The answer is that we don't expect virtual worlds to be any more absorbing than any of the technologies that preceded them. Books can be highly immersive, and we all know people we would categorize as bookworms, who prefer staying at home and reading a good novel over going clubbing, and serious Netflix buffs who, say, binge-watch a full series of a show over a single weekend.

Both types exist without losing their interest in, and ability to function in, the physical world – you may be one or both of those types yourself.

But it is true that the virtual world industry itself can play a conscious role in where it encourages Virtual Natives to be, and what it encourages them to do. For example, Niantic CEO John Hanke has expressed his hope "that the world doesn't devolve into the kind of place that drives sci-fi heroes to escape into virtual ones."[8] Instead of VNs losing themselves in that Shrek-adorned version of the immersive virtual Met we conjured earlier, Hanke's vision is one in which our digital worlds "give us a reason to call a friend, make plans with our families, or even discover brand new friends." By bringing the unlimited imaginative potential of virtual worlds into our physical space through augmented reality tools, he asks, "Could [this technology] help us discover the magic, history, and beauty hiding in plain sight?" His answer is clearly yes.

And when people get too fanatical online, it's useful to remember that a current phrase in Gen Z slang is, "touch grass." This is an insult that means it's time to get off the computer and go outside, and is usually directed at people who seem to be a little too deeply involved with their online worlds. In other words, the Virtual Natives are looking out for themselves, and each other.

From Apprentice to Master

We began this chapter looking at Virtual Natives as digital experts, then examined the pervasive fear that this is a generation that is acquiring that expertise at the expense of a rich life lived offline. Our view is that time spent acquiring expertise in any field requires just that, time, and there will always be a trade-off between what a Virtual Native – or anyone – is doing and learning right now and what they *could* be doing and learning. But is that any different than the world has ever been?

In some ways, the expertise in virtualization that VNs are building may be akin to the guilds or apprenticeships of yesteryear, in which adolescents dedicated themselves to a task for years as part of the accepted process of mastering it. Today's Virtual Natives are not only learning a trade, but they are also developing the skills they need to successfully run the business of themselves, skills that they possess in greater measure and at a younger age than any generation that has gone before them.

The digital fluency of Virtual Natives is marvelously demonstrated in the 2023 Netflix film *Missing*, which takes place entirely on the various screens in the life of fictional teenager June as she searches for her mysteriously vanished mother. In the film, June skates smoothly and confidently from program to program and device to device, unlocking the secrets of her mother's past through online sources such as browsing histories and security camera footage. Just like June, Virtual Natives are maestros, who know how to call upon every instrument in the digital orchestra available to them.

Mindset makes all the difference. The time spent by Virtual Natives in virtual worlds, or using digital tools, is not time wasted, but an investment in the development of their wide-ranging technical expertise that sets them apart as perhaps one of the most useful generations of all. They are putting in their apprenticeship years on the way to mastery. The task for those of us who are older is to recognize this expertise for what it is, honor the time that went into developing it, and figure out how we can leverage VNs' hard-won digital knowledge to improve both the digital and physical worlds.

7

Create, Consume, and Own

BEFORE THE ADVENT of cable television, television was programmed. You had to be at home at a certain time to watch an episode, or risk missing the show, perhaps forever. And the programs that played were generally wholesome fare, designed to appeal to as many households across the country as possible, since everything was broadcast nationally on one of the three major channels. *The Brady Bunch. Gilligan's Island. The Waltons.* Television was pleasant and unchallenging.

This began to change in New York City in 1971, when the introduction of cable TV there brought with it two public-access channels dedicated to "commercial-free self-expression."[1] Cable TV wasn't subject to the decency rules that limited what could be shown on regular broadcast TV, so during New York's public-access golden era, the 1970s and '80s, anyone could turn up at the cable studios and broadcast, well, anything.

And they did. One example of "anything" was the videos created by a man named "Ugly George," who scoured the mean streets of Manhattan wearing a silver cut-out top and shorts and a full camera rig on his back. He sought and convinced aspiring actresses to duck into New York City stairwells and undress for him while he captured them on film, videos he would later share on his Manhattan cable show.

Robin Byrd, another early cable pioneer, hosted a talk show. On any given night her set would be aflutter with adult porn actors, dancers, male go-go boys, dominatrices, or their "slaves," bedecked variously in feathers, glitter, cowboy outfits, studs, or mesh – and sometimes, nothing at all. By 1993, New York boasted 70 cable channels with what was considered "the greatest concentration of producers, network executives and eager-eyed video junkies of any place on the planet."[2] This flowering of self-expression and boundary-pushing was one of the first indications that when a platform is created that allows anyone to broadcast themselves and their worldview for free, there will be no shortage of people ready to step into that spotlight.

What very few could have predicted in New York City in 1979 was that 40 years later, videos and other content made by ordinary people, the spiritual (and generally much less sleazy) descendants of those early public-access adventurers, would be the dominant media created and consumed by the Virtual Native generation, many of whom watch broadcast media rarely, if ever at all. Happily for them, this content creation is accompanied by a key feature that largely eluded those early cable experimenters: the ability to make serious money.

Virtual Natives grew up first on YouTube and then later on Instagram and TikTok, all platforms that encourage each and every VN to create their own channel and to be the star of their own show. From unboxing videos hosted by chirpy five-year-olds to the millions of dollars raked in by teenage makeup influencers, the ease with which VNs can generate, share, and now even monetize their own messages has given this generation both an unprecedentedly wide worldview, and, most important, an ability to generate and allocate their income at a young age that has never been seen before.

Instead of buying into the hamster wheel of fate "commute + dedicate yourself to the company" (sleep, wake, commute, work) ethos that previous generations have unquestioningly accepted, Virtual Natives are abstracting work down to essentials, i.e. "do function = earn money." This leaves them free to write their own script about when and where they do that work, how long they want to stay with a company (not long), and how picky they want to be about the ethics of the company they join (very). Or literally leaves them time to write actual scripts.

This mindset is another outcome of the financial turmoil that VNs have seen Millennials go through, which we mentioned earlier. In 2022, for example, McKinsey found that less than half of Gen Z thought they'd ever be able to buy a house.[3] With these bleak economic expectations, it's no wonder that Virtual Natives are not particularly excited about sacrificing their lives to an employer who is likely to fire them at the first sign of market trouble.

Given all this, many Virtual Natives are not looking to the traditional workplace as a possible career source at all. Even more significant for understanding the money-earning psyche of the Virtual Native is to understand new ways of making money that have nothing to do with office jobs, and everything to do with developing a fanbase.

From Hobbies to Full-time Hustle

When Priyanka turned 13, the most exciting thing that she did on that milestone day was to create a Twitch account for herself, because that was the first day she was old enough to do so. And she had incentive. If a Twitch audience likes a gamer's style, they can follow the gamer as a subscriber, which generates an income for both Twitch and the gamers themselves.[4]

Priyanka started broadcasting video of herself playing her favorite games and narrating her exploits. It's a long slog to the top, though, and she eventually moved on to other dreams, but plenty of other people use Twitch as a source for their full-time income. It has been estimated that expert and charismatic players who stream on Twitch for 40 hours a week or more can earn $4,000 per month, and even more with ad revenues and sponsorships.[5] That's pretty good money, especially considering that you can start earning this kind of income at the age of 13, and you do it while playing the games that you love.

The Twitch business model is notably different from the way that influencers on Instagram earn their money, which is primarily generated by advertising and sponsorship. Twitch's "I will subscribe to your feed and pay you if I like your content" model allows fans to directly express their appreciation and support for their favorite gamers, which in turn allows the gamers to make a living from doing something that doesn't feel like work at all.

It's the Virtual Native version of going to watch a professional sports match, except that instead of paying an entrance fee that goes to the team owner, you enter the stadium for free, then walk up to the players after the game and just hand them some cash if you like how they played.

Even though Twitch is owned by Amazon, which, of course, gets a cut of all the money that flows from fans to admired gamers, the overall mechanism is still one that largely defies traditional notions of what work is, and how, when, where, and why it should be rewarded. It's such a successful model that we see it being repeated in multiple platforms that appeal to Virtual Natives, such as the music sharing smartphone app Encore.

Looking for a way to reconnect musicians with their fans during Covid, rapper Kid Cudi created Encore, which received US$9 million in funding in 2022.[6] Encore lets musicians create their own visually exciting "concerts" using Augmented Reality that then stream on the app. Fans can reward their favorites with "Claps" that cost 10 cents each. The artist gets eight cents, the platform gets two. The fans who clap (and therefore pay) the most earn exciting perks such as winning merchandise or being able to meet their favorite musicians online.

Platforms like Twitch and Encore, where Virtual Natives currently spend massive numbers of hours during their preteen and teenage years, are teaching Virtual Natives that you don't have to "go to work" to earn a living. In fact, the digitization and monetization of desirable creative content – whether an eSports match, an Instagram unboxing video, or a great song – is fundamentally enabling lifestyles based on creative pursuits. These platforms and business models are creating genuine alternatives to the academic path that can lead straight to the office cubicle that was more familiar to earlier generations. And in fact, by the time the average Virtual Native graduates from high school, chances are that they have been exposed to FAR more examples of people earning money online than they have been exposed to more traditional workplaces. Take Your Daughter to Work Day is once a year; "Watch Instagram Influencers Rake in the Cash" happens all 365!

Ryan Hilliard, of HypeAuditor, a company that analyzes data associated with online influencers, estimates that it takes about one million followers for an influencer to break through and earn enough money through sponsorships, gifted merchandise, and direct product promotion to make a full-time living.[7] He also estimates that about 1% of would-be influencers actually make it to this level. But the ones who do make it are so visible that new wannabes are continually inspired to try to get there, too.

Onyx Clark, 25, who is a teacher, is trying to break into the influencer biz as a video game reviewer and streamer. She captures what drives her to have a side hustle this way: "My family are immigrants. I'm the only one born here in the U.S. So a lot of them are just like, what are you doing? But I feel like not taking a chance on yourself is really the thing that'll like – that's worse, honestly."[8] And this is what motivates so many VNs to become content creators – the opportunity to make money online doing something you love is . . . just . . . so . . . close . . . that they'll never forgive themselves if they don't at least give it a shot.

Another source of online earnings available to VNs is through gaming. *Fortnite*, for example, hosts multiple tournaments open to all players good enough to qualify, and the winners earn serious money. In 2019, 16-year-old Kyle "Bugha" Giersdorf earned $3 million by becoming the Fortnite World Cup champion, inspiring teenagers around the world to attempt to emulate his feat.[9] But players don't have to win the really big tournaments to earn enough money to change their lives. Ismail, 16, whose mother works as an office cleaner, has cleared more than $36,000 through his *Fortnite* victories, enough money to make it far easier for him to attend college and change the likely trajectory of his life.

Free Agents

Virtual Natives are not just at home on online platforms; they know how to massage them to generate income.

In 2018, Montero Hill was couch surfing. Which is to say, he didn't have an apartment of his own and was coasting on the sofa, and goodwill, of his brother, after being tossed out of his sister's house. He had dropped out of college. His Wells Fargo bank account was negative $5.62. But despite these challenges, Montero had a special skill that many lacked; he knew how to go viral.

As @nasmaraj on Twitter, Montero mastered the art of the Tweet Deck – a now-banned practice in which networks of popular accounts coordinated to "game" the system by agreeing to retweet and promote each other – and co-opt viral tweets from less prominent accounts.[10] Through crafty Twitter manipulation, Montero was able to garner over 10,000 retweets of his posts. Montero was an aspiring musician, and to get to the top, he suspected he could leverage these same skills to defeat the algorithms and get his work noticed. He came up with the idea of a crossover, genre-defying music beat that would get him noticed quickly in a small pond, so that he could then cross over to the bigger pond of hip-hop.

In December 2018, Montero tweeted simply:

"country music is evolving"[11]

He had spliced his song, "Old Town Road," over a hand-shot video of a guy in a cowboy hat at a rodeo dancing uninhibitedly. It was reposted over 30,000 times. Montero then cross-promoted it on Reddit, Twitter, Instagram, and YouTube, where he matched his music to a video montage of cowboys riding their horses and doing cowboy things, lifted straight from the video game *Red Dead Redemption 2*. Soon, videos of people dancing to the tune went viral. The song went on to be the number one song on the Billboard US Country music charts – until the industry realized who was behind it and how he got there. Montero was pulled from the charts.

Country music star John Rich appeared on Fox News to discuss the incident, saying, "I don't like people that try to piggyback on real country music. If you really want to be a country artist, then be one – come to Nashville, write your music, really come up with something that's fitting somewhere around country music."[12] In other words, there was a process that artists were expected to follow, and what Montero had done wasn't it.

Undefeated, Montero – now known as Lil Nas X – reverted once again to social media. He shared a video of himself listening to "Old Town Road" and posted:

"Twitter please help me get billy ray cyrus on this"[13]

And Billy Ray Cyrus was game. Only a few months after Lil Nas X's original post, Cyrus joined him in a duet re-release of "Old Town Road" that went viral. The two went on to perform the song at the 2020 Grammy Awards, where they won "Best Music Video" of the year for a charming rendition of the song that definitely did not include bootleg *Red Dead Redemption 2* footage.

In under 12 months, a broke, homeless, unemployed teenager in Atlanta was able to launch a music track, remix it with a country music legend, and break a record in Billboard Hot 100s 60-year history by charting a country tune at number one for 19 consecutive weeks. All because he understood algorithms, content creation, and the power of digital platforms.

Digital Goods, Real Incomes

Okay, so not every Virtual Native is going to be able to write and release a song that goes to the top of the charts. But with access to the same digital tools and platforms that Lil Nas X used, the average VN does have the opportunity to transform their artistic skills into earnings.

And they do have artistic skills, in spades. A recent study commissioned by Snap Inc. found that 80% of Gen Z respondents classified themselves as creative, a significantly higher proportion than earlier generations. (In the same survey, 90% also said they considered themselves and their friends to be kind.)[14] Seventy-seven percent reported in a different poll that they pursued creative pursuits offline as well as online, in areas such as drawing, writing, or playing a musical instrument.[15]

In addition to the money-generating options open to VNs that we've already looked at – becoming an Instagram influencer, Twitch streamer, or *Fortnite* gamer – even more VNs have created digital goods for sale in fashion-focused platforms like Roblox and ZEPETO.

ZEPETO, a popular Metaverse platform that sells virtual items of high fashion from brands like Christian Dior, Givenchy, and Ralph Lauren, now boasts over a quarter of a billion users, 90% of whom are located outside of its South Korea home base. As with Roblox in the United States, brand-name digital clothing is only part of the story; ZEPETO is also extremely welcoming of independent creators and sellers. ZEPETO Studio currently has 2.3 million creators designing and selling their digital fashions on the platform. Roblox boasts even more, with 11.5 million creators selling 62 million items in 2022, a 25% increase over 2021 sales volumes.[16] Both platforms allow creators ample scope to develop and promote their own brand, with Roblox even introducing Verified Badges in 2022 so that well-known independent designers can ensure that their productions are indeed original creations and not copycat knockoffs. In fact, Roblox estimates that there are around 200 times more digital fashion designers actively working in Roblox than there are creators of fashion collections out in the physical world.

There is a strong market for all these digital fashion creations. Three out of four Roblox users spend money on digital fashion, with 25% confessing that they have spent more than $20 on a single virtual item of clothing. Half of Gen Z Roblox players change avatar clothes every week, with 53% reporting that they will change their avatar's look to suit their mood or feelings that day.[17] While selling outfits within Roblox isn't a guaranteed path to riches for all, a topic to which we will return later, it is still an undisputedly powerful forum for learning how to design and sell digital clothing, and there is social cachet in becoming known for your designs. Fifty-eight percent of Gen Z Roblox players agree that being a digital fashion designer is just as impressive, or even more so, than being a designer of physical fashion. To Virtual Natives, digital platforms are a legitimate source of income, and a viable career choice.

Even demographic segments that you might not associate with being interested in fashion, like teenage boys, are making money from and spending money on how they look online. In May 2023, British Mum @UjuAnya tweeted, "A laptop arrived at my house addressed to my 16yo son. I didn't buy it . . . I sat the boy down [and told him I'd] . . . ground him indefinitely until he said where he got $1000. [He told me the money came from selling] Video game skins."[18] Other Twitter users didn't even know what video game skins were, and asked her to explain. Turns out she hadn't known what they were, either. "Bruh. I grilled him like the CIA. From what I understand, they're digital appearance modifications for video game characters . . . something like Barbie doll clothes and accessories. You buy and trade the skins with other players inside the game."[19] Exactly right. And it's in that digital goods marketplace for people who want to look a certain way online that a 16-year-old boy can make $1,000 – legally. Many respondents on @UjuAnya's thread were also astonished to learn about this digital goods market and her son's prowess in it, with one non-VN commenting, "Even being 25 in this world now feels like 50:/."[20] Notice as well that @UjuAnya's son plowed his earnings right back into his computer infrastructure, rather than blowing it on candy. He's investing in his business.

While we're on the subject of digital fashion, let's take a quick detour and compare how Virtual Natives dress online with how they dress in the physical world. At the same time that VNs are elegantly outfitting their online avatars with Gucci and Balenciaga, they're wearing almost aggressively comfortable clothing – pajamas, in fact – to school and work. A 2021 *Wall Street Journal* article captured the zeitgeist in an article entitled "Pajamas Are the New Sweatpants,"[21] and a look into the hallways of any high school in the United States will confirm that both male and female VNs are wearing fuzzy pajama bottoms and even bedroom slippers out in public on a regular basis.

When asked about the disparity between his exquisitely dressed *Fortnite* avatar and his more comfy – and slouchy – clothing choice for his physical body, 18-year-old Luca put it this way: "The way I dress online is. . . I guess it's aspirational. It's what I'd like to be. But in real life, you have to compromise. I can't afford to buy any of that stuff my avatar wears. I might as well be comfortable, so here we are." ZEPETO designer Monica Louise agrees. "There are clothes that I cannot afford to wear in real life but in the digital world, I can buy all of them," she affirms.[22]

Virtual Sweatshops?

Most of the platforms with digital goods that Virtual Natives frequent – ZEPETO, Roblox, Minecraft – are aware that their main user base is children. They all have digital marketplaces, but none of them allow the sale of Blockchain-based non-fungible tokens (NFTs), in order to avoid the speculation and even fraud that is associated with NFT creation and trade.[23]

As VNs age out of childhood, though, they may well shift their creative skills in the digital arena away from gaming platforms and into NFT creation, because some of the gaming platforms do have a problem: adequately compensating the children who design for them.

Taking the example of Roblox, there are many ways that anyone can make money on the platform. In addition to offering sales opportunities for digital goods created by its players, Roblox encourages kids to earn money through the development of new games and worlds. In Roblox Studio, you learn how to build a game from scratch and test it. Creators can monetize their games to earn revenue, both by charging people to play their games and by offering pay-as-you-go in-game purchases.

These moneymaking opportunities have led to some notable success stories. Identical twins Ben and Matt Horton began playing Roblox when they were 10 years old. Just three years later, their hobby changed the fortunes of their entire family when their game "Boat Ride" became a hit. At 13 years old, the boys were earning a "significant" amount of money. Today, at 21, they are both developing games full-time, earning healthy six-figure salaries, and have even paid off their parents' mortgage.[24] Statista reports that in 2022, 774 Roblox developers and creators earned $100,000 or more, with 99 of those bringing in more than $1 million.[25]

But it turns out making actual money in the game is not that easy for the vast majority of creators. The same Statista report that shows that 774 developers earned $100k or more in 2022 indicates that 4.2 million developers and creators earned *nothing at all* from their Roblox activities that year. The developers earning six figures and more make up .018% of the developer community, a vanishingly small proportion, and a tiny fraction of the 1% who strike it lucky as influencers on Instagram or TikTok.

Take Jack, for example. Jack created a hit Roblox game in 2021 when he was just 13 years old, earning him about $700. He gleefully spent roughly half of his earnings on in-game purchases, but when he tried to cash out the remainder, the game wouldn't allow him.[26] Roblox has come under serious scrutiny not only for charging developers 30% of all in-game game earnings, but also making redemptions so difficult as to be nearly impossible to cash out.

Because 67% of Roblox players are under the age of 16 (and only 14% are over the age of 25), most of the creators of both games and goods on the platform are indeed children.[27]

This, plus increasing publicity about child builders not being fairly compensated for their successful creations, has led to articles with headlines like "The Trouble with Roblox, the Video Game Empire Built on Child Labour," which appeared in the *Guardian* in 2022.[28]

In a scathing pair of videos on YouTube, People Make Games journalist Quintin Smith broke down the numbers and revealed just how hard it is for Roblox developers to wrest any earnings from the platform.[29] In a subsequent podcast interview, he likened Roblox to an exploitative "company town" in which "the workers are kids, and the kids are literally being subjected to crunch and burnout," as he explained in a podcast interview.[30] He calculated that after taking all of the platform's rules and restrictions about redeeming money, the children's labor was rewarded under 17 cents on the US dollar. In summarizing his findings about the skewed system, Smith declared, "It would be illegal if it wasn't happening online."

In the physical world, employing children under the age of 15 is difficult in most states, and even then they can only work certain kinds of jobs. Engaging child labor for free is unethical at best, and illegal at worst. Yet, in game worlds, kids genuinely enjoy designing, building, and making their creations available to others. Moreover, having children design games for children is a bonus; they know what other kids will like. If you're Ryan Kaji, unboxing toys on YouTube in exchange for millions of dollars in sponsorships and endorsements, that's one thing. If you're a child game designer, spending untold hours building, advertising, and selling games, to receive less than one-fifth of the money that you've earned, that's another issue entirely.

Luckily, Virtual Natives are savvy, and critically, they communicate. They do have skillsets and the entrepreneurial mindset to take their games and build elsewhere.

In its current state, Web 2.0 is widely seen to have been hijacked by large, profit-seeking platforms. This may explain the impulse toward the equitable as one of the driving forces behind

Web3, the next iteration of the internet. It consciously tries to recapture the democratic, decentralized ethos that informed the original incarnation of the World Wide Web.

Web3, which focuses on DeFi, or Decentralized Finance (and which sounds perhaps not coincidentally like the word "defy"), uses Blockchain to redistribute ownership of resources away from large platforms and back into the hands of the general community of participants and builders. It remains to be seen whether this goal will be attainable (what's to stop the next wave of controlling platforms from arising, this time perhaps in the form of crypto wallets, or other structures monopolizing key elements of the Web3 world?), but Web3's determined rebellion against those who make and manipulate the rules is a significant feature of Virtual Native thinking.

Which brings us back to NFTs. One of the reasons that adult visual artists around the globe have so wholeheartedly embraced NFTs as a way of releasing and selling their images and video clips is that the Blockchain keeps an eternal record of who originally created the work of art. And – here's the important part – that original artist gets a cut of the sales price every single time that work of art changes hands in the future. When Van Gogh's *Irises* sold at auction in 1987 for $53.9 million, did Van Gogh's estate see any of that money? Absolutely not. But thanks to NFTs, the descendants of today's digital Van Goghs will benefit from future sales of their creative ancestors' work forever. And even more to the point, this eternal revenue structure for digital art finally gives creative producers of all kinds a path toward earning a steady living, a challenge that has been much more difficult for many of them in the past.

While they're still young, and designing for fun, most VNs are happy to stay on platforms like Roblox and ZEPETO that give them easy-to-use tools and a large, ready-made audience of potential buyers and fans. But as they get older, and start to wise up to the potentially exploitative elements of some of these platforms, it is possible that Virtual Natives will take matters into their own hands and depart for more equitable blockchain-based arenas, or at least forums that give a better return on VNs' investment of their time and creative juices.

Even though not all platforms on which Virtual Natives can potentially earn money are fair ones, the central point remains: Thanks to digital access, Virtual Natives have the potential to legitimately generate an income while still at quite a young age. The significance of this is profound. What does it mean for millions of members of an entire generation to be actively compensated for their skills before they've left high school, or even middle school? Among other things, they learn that they have agency, and their output has value. This creates a mindset in which they might not be quite as grateful or even obsequious to potential employers as previous generations have been, a topic that we will explore in the next chapter.

8

Autonomous Agents

WE'VE SEEN THAT Virtual Natives have been encouraged from a young age to take action for themselves on digital platforms, and that some of those actions can be financially rewarding for them. Digital platforms also enable VNs to create physical businesses. Fourteen-year-old Eva has found that "things that look difficult – like starting my . . . online jewelry-making business – actually aren't that hard if you get into it."[1] Learning early in life that one has agency and can make decisions for oneself is empowering. Eva continues, "I've realized that I can try for one career, but if it doesn't work out, there are always other options, other opportunities." This is a piece of wisdom that Virtual Natives know far better than many of their elders.

Gen Z is eager to embrace innovation, and has no illusions about the risks involved in embracing truly bold new ideas. A recent survey found that more than 80% of Gen Z respondents agree that "embracing failure on a project will help them to be more innovative," correctly spotting that fear of failure is one of the great inhibitors of risk-taking.[2] In the minds of Virtual Natives, there is no failure, only an opportunity to learn how to do it better the next time if things don't go to plan.

At the same time, the prevalence of gaming among Virtual Natives has conditioned many of them to expect to be able to "level up" in some aspects of life. One Millennial boss found himself telling an impatient VN underling to slow down, "It's not your time yet," a sentiment that made sense to the older worker, but not to the younger one. The VN responded, "Okay, what mile am I on? How fast am I going and what do I need to do to go faster? . . . Give me specific milestones. . . Give me specific skills to build. Tell me exactly what I need to do to get there."[3] In life, as in gaming, Virtual Natives want to know what it takes to reach the next level, which may not always mesh with the slower, incremental career progress associated with many traditional workplaces.

We've already mentioned the powerful influence that the economy of the last few decades has had on Virtual Natives and the life choices that they make. Because this topic gains additional relevance when discussing VN attitudes toward their employment and careers, and the control that they desire to have over both, let's take a deeper dive into the exact marketplace headwinds VNs face.

No More Bullshit Jobs

As one Virtual Native puts it, "The definition of 'success' has changed. My boomer folks lived on one teacher's salary, bought a home and cars, raised three children, had no student loan debt, and retired at 65. That way of life is no longer available to future generations."[4] How did we get here?

In 1997, Alan Greenspan, the Head of the US Federal Reserve, faced an economy with historically low unemployment numbers. Ordinarily, low unemployment leads to wage inflation as companies vie for talent by offering better salaries, and the Federal Reserve tries to head off this inflation by raising interest rates. But this time, Greenspan was reluctant to raise rates. He explained his hesitancy by saying that while "the softness in compensation growth," meaning low pay, was remarkable, he "would be surprised if it were nearly as important as job insecurity."[5]

Greenspan's point – and the reason he gave for delaying an increase in interest rates – was that workers didn't organize as if they were in control. Why not? Job insecurity. By being constantly fearful of losing their jobs, workers were less confident to push for wage increases.

Since then, job insecurity has continued to plague American workers. A 2016 article by the World Economic Forum suggests that the global transformation in labor relations that began with market liberalization in the 1980s – and the concomitant shift to a highly financialized, postindustrial economy – has led to increased corporate influence, lower taxation rates on both large companies and wealthy individuals, and a reduction of collective bargaining power.[6] Having the upper hand, corporations have been able to set employment terms that favor them, often granting employees little or no control over key lifestyle factors such as when their shifts are scheduled or how many sick days they are allowed.

Hustle Harder: The Birth of the Gig Economy

To meet the needs of those who could not make their lives conform to the strict requirements of larger corporations, a host of labor "flexibility" options introduced by start-ups and institutions alike began to offer unstable jobs with no benefits, health insurance, vacation time, or holiday pay. The "gig" economy was born.

The rationale for "gig" jobs was that they provided workers alternatives to full unemployment, and offered flexible work schedules that would enable staff the freedom to pursue other forms of employment or attend to personal interests such as child or parental care, education, or vocational training. This sounds great on paper, but the reality for people's actual lives has been very different.

Author Guy Standing uses the term "the precariat" to refer to the perpetually underemployed, those with eternal partial employment. He notes that this state "creates existential insecurity, and goes with the fact that for the first time in history many people have education above the level of labour they can expect to obtain."[7] The "precariat" describes a new generation of working-class people, similar to the working-class proletariat, but whose lives are made "precarious" because their always-shifting employment situation provides them with very little or no financial stability. Their short-term jobs increase economic volatility and increase the precariat's prospects of downward mobility with low likelihood of recovery. Even more important, the precariat suffers psychologically as well, with "no occupational identity or narrative to give to their lives."

Thomas Piketty, professor at the Paris School of Economics, points out that "the democratic system is not enabled to have a common-sense reaction to this excessive level of inequality that, in the long run, is not good for U.S. prosperity."[8] According to the Federal Reserve, about 64.3 million – or half – of American families own just 1% of the nation's wealth.[9] Says Piketty, the inequality gulf in the United States is "probably higher than in any other society at any time in the past, anywhere in the world."[10]

Unsurprisingly, Virtual Natives – generations born in the dawn of the 2000's – have never known an era of relative prosperity, and have grown up with perpetual economic insecurity.

Both Gen Z and Gen Alpha grew up in a period of negative income growth, with the lowest quintiles feeling the pinch more than most. They are aware that previous generations, notably Boomers, grew up in an era of relative economic prosperity across the population, with incomes of the lowest fifth quintile bucking the historical trend and, unusually, rising faster than the highest economic brackets. It was a time of hope and growth for all.

Okay, Boomer

"Breakfast? What? When I was your age, I woke up before dawn to sell newspapers for a nickel apiece, then I walked to school uphill, through the snow, barefoot – both ways. I didn't have time for breakfast!"

"Okay, Boomer."

The phrase "Okay, Boomer" (a popular Gen Z insult intended to point out an anachronism uttered by older generations) evolved from the disconnect between the widely divergent economic realities for each generation, and their relative socioeconomic advantages. While the Boomer generation benefited from broad government policies favoring education and income growth for the working class, that is no longer the case. The onus has since been shifted to workers who are encouraged to be more resourceful and work harder while being constantly admonished to cut back on expensive caffe lattes and avocado toast.

But it's not about the lattes. "The key reason the U.S. economy was so productive historically in the middle of the 20th century was because of a huge educational advance over Europe,"[11] says Thomas Piketty, citing that, in the 1950s, high school attendance was as high as 90% in the United States.

Since the 1960s, much has changed in the structural tools traditionally used for economic advancement. Take college, for example. Historically, holding a college degree advances economic opportunity. US adults with a bachelor's degree typically earn an average of 75% more over their lifetimes than they would have with only a high school diploma.[12]

However, the cost of obtaining a college degree has, for many US students in recent years, become increasingly out of reach.

The average cost of college has more than doubled in the twenty-first century.[13] The average private, nonprofit university student spends a total of $54,501 per academic year, 70% of which is tuition and fees. Additionally, studies by the Center on Budget and Policy Priorities shows that funding decreases since the 1980s have had a negative overall impact on the quality of education in the United States, as colleges are repeatedly forced to reduce faculty, limit course offerings, and reduce student scholarships and grants.[14]

Late-Stage Capitalism and the Revenge of the Precariat

Economic instability, a widening inequality gap, and a greater debt burden weigh heavily on Virtual Natives.

A survey by Instagram in 2022 found that nearly 90% of Gen Z felt that "too many people are forced to work multiple jobs to make ends meet"[15] while *Forbes* magazine estimated that same year that there were some 735 billionaires in America whose collective worth was over $4.7 trillion.[16]

Indeed, the United States has a uniquely large underclass of people who spend their lives cycling through low-paid, dead-end jobs, rarely accumulating any savings at all. In data stretching back to the 1990s, about a quarter of US workers are classified as having low pay.

Compared to their developed world peers, Americans face higher job insecurity and precariousness, which stubbornly persists, despite evidence that the negative effects of job insecurity on workers' physical and mental health is detrimental to both workers and their employers.[17]

Research also reveals that when pressured, job-insecure workers are more likely to adhere to the minimum job requirements, but not more.[18] Studies by Dr. Mindy Shoss show that people who are worried about losing their jobs, or who work in temporary positions as contract or "gig" workers, prioritize visible contributions, like turning up on time to meetings, or responding rapidly to important internal or client emails. That is, threatened workers focus on performative effort, rather than actual performance. While their contributions are easy to see, they are not necessarily adding value to the organization.

In recent years, we've developed a term for this kind of behavior: quiet quitting. And young people are doing it the most.

Quiet Quitting and the Rise of Alternative Employment

TikTok star Sarai Soto makes viral (and very funny) videos about toxic workplace conditions, calling out ludicrous expectations of Millennial, Gen X, and Boomer bosses. She popularized the phrase, "Act your wage," which means putting in an amount of effort that matches how much you are being paid to do the job. It's not about doing a bad job or being eternally disengaged. "It means you do your best without being manipulated into doing more than you're being paid for and burning out."[19] To a fictional boss who wants her to go above and beyond the work that she was hired to do, she retorts, "I'll work more when you pay me more."

Similarly, Deandre Brown, aka "The Corporate Baddie" regularly shows Gen Z how to prevent fictional boss "Jack" from crossing boundaries. When Jack oversteps them – by asking him to work late or deliver more than required – he explains why that won't be happening. "Working Sunday nights is not in my job description, Jack."[20] In other videos, when Deandre sees emails appear on a Friday at 4:54 pm, he simply closes his computer and walks away.

To Virtual Natives, this approach is about self-preservation. To bemused older generations, who have slaved to the bone to exceed their managers' expectations in hope of future advancement, this looks like "quiet quitting," with an emphasis on "quitting." It's puzzling behavior, and borderline unthinkable.

But not to Virtual Natives. "Just because [older people] had to suck up their problems and carry on doesn't mean we have to. Sometimes, I feel like not overworking ourselves to prioritize our mental health is often mistaken as laziness, which isn't the case," commented one VN in early 2023. Another feels that "I won't prioritize work over my mental health or personal time. If any company tries to pull that, I'll quit in a second."[21]

Part of what underlies this VN determination to not be crushed by their employers is their front-row seat for what happened to their Millennial and GenX parents, who did jump through all the hoops at work, but still got laid off and had to struggle to keep up with their mortgage payments. Another contributing factor is the agency and sense of control that they have developed through their powerful command of digital tools – tools that give them options.

Why flip burgers when you can earn a passive income as an influencer, with a marketplace on Amazon, selling virtual clothes, trading NFTs, or posting pictures of your feet on OnlyFans?

VNs aren't just dancing on TikTok and Reels; they're planning for the future. Popular themes on social media include wealth generation, entrepreneurship, and creating a passive income.

YouTuber Jasmine McCall, for example, has reported making an average $105,000 per *month* in passive income from, appropriately, her wealth creation advice videos. What started as a fun side hustle became so lucrative after an early video explaining how she improved her credit score went viral that she quit her $124K annual pay job at Amazon and focused full-time on her YouTube content – for which she only needs to work two hours per day.[22]

With countless stories like Jasmine's circulating on the internet, VNs know that a secure future doesn't necessarily depend on retaining traditional employment. Those who hold jobs no longer sacrifice everything for their employers, citing the outsized expectation of workers to commit their lives to jobs that give little commitment back to them beyond basic compensation. Virtual Natives recognize that the relationship between employer and employee is purely contractual, and that the contract can be severed at any time. They quit without notice. They close their computers at 5 p.m. They leave when they don't feel their efforts are valued. They are reclaiming their value, and their time.

"Individualism is a massive thing for us," says Michael, 21. Because there are no immutable rules, he says, "my generation is 100% more entrepreneurial. . . We want to make stuff happen for ourselves."[23]

The very job insecurity that made previous generations more loyal, according to Greenspan in the 1990s, is now backfiring. Virtual Natives have options, and they are exercising them en masse.

Virtual Natives, Virtual Work

Cooper, Khendra, Erin, and Hadley were brainstorming their working futures. But instead of listing industries that interested them or corporations they might want to join, they huddled around a whiteboard, where they had written out some questions:

- What's your side hustle?
- What platforms are you most active on?
- Where is your biggest follower base, and what content do you create?
- If you had $10 million in VC money, what would your start-up be and how would you launch it?

With Virtual Natives most naturally thinking of their digital skills and fanbases when planning their careers, it's no surprise that Human Resources departments have a tough time matching today's abilities with yesterday's job requirements. Virtual Natives are hustlers. Between gigs, or part-time work to stay afloat, and "side hustles" in which they pursue passion projects, and with their high tolerance of the ramifications of failure, VNs are entre-preneurs, constantly seeking ways to outperform the competi-tion. Whether it's measuring their own social media performance against peers, managing their investment portfolios, crypto wal-lets, developing virtual assets and more, VNs are actively pursu-ing their own careers, and on their own terms.

Stacy, for example, is studying investment banking at univer-sity, but her side hustle is fashion content creation for TikTok. Even though the two fields aren't directly related, she says that her fashion work gives her an edge in her banking outlook. "Cre-ating content for brands and working with marketing teams, it's allowed me to have a different perspective and be more unique and creative in how I approach things. . . I'm definitely continu-ing my side hustle post-grad."[24]

Ali, another university student, has a different side hustle, one that earns her five digits annually. She looks for good-quality clothes at local thrift stores, then resells them on apps like Poshmark and eBay. This put her in a strong position to continue earning money when the pandemic hit. "I was super-fortunate, because if I [had had] a more conventional job, I probably would have been let go."[25] Hustling on digital platforms is a natural way for Virtual Natives to retain control over their own destiny.

The hustle mentality is a far cry from the "loyal employee for life" mindset of earlier times. A 2022 study found that Gen Z was changing jobs at a rate 134% higher than they were in 2019.[26] In comparison, the same study found that Millennials were switching jobs only 24% more, and Boomers 4% *less*.

Ask a Virtual Native what they think is a reasonable tenure to stay in a job today, and many will respond that "about a year" is a reasonable time to remain in a given role. In fact, the current average for Gen Z to hold a job is two years and three months, the shortest period of any generation.[27]

Twenty-one-year-old Arial Robinson explains that she once liked the idea of being loyal to her employer, but experience in the working world quickly make her rethink her priorities. "I feel like these companies don't really have loyalty to us . . . at the end of the day, they're trying to make money and run a business and they will get rid of us just as quickly as we could get rid of them."[28]

Right.

Employment is a mutually agreed contract between contractor and contractee. But the balance of power is shifting.

For decades, employers had the upper hand, keeping most employees loyal due to scarcity of employment and cultural norms that convinced people to demonstrate loyalty to the companies where they worked. That no longer obtains.

In 2022, 65% of Gen Zers reported that they planned to leave their job within six months.[29] That represents over 40% of overall workers, which suggests that a great "reshuffling" is taking place. But why?

The first motive is fairly obvious: money. Nearly 40% of study respondents claimed that salary and/or potential bonuses were their primary reasons for wanting to leave their jobs. In 2023, Adobe's Future Workforce Survey[30] found that 85% of upcoming and recent graduates would not apply for a job if the salary was not listed. And as we saw earlier with Jasmine McCall, visible examples of successful content creators abound. Eric Sedeño, who streams on TikTok as @ricotaquito, left his corporate job as an art director at an advertising agency and is now earning "more money" as an online creator through brand sponsorships, largely thanks to his warm, authentic personality.[31] For VNs who see these success stories in their feeds on a daily basis, the lure to do the same is powerful.

The second major motive, perhaps surprising to older generations, is learning. What can VNs learn on the job they can't get from TikTok or YouTube? This is actually a serious question. When employees spend more time working from home, what opportunities are companies offering for upskilling and development that VNs aren't already learning from YouTube University, or peer discussions in social media?

In addition to salary and opportunities for skill growth, a third factor of importance to Virtual Natives is ideals. A 2022 Employee Retention study[32] found that, even more than money, 42% of Gen Z would rather be at a company that gives them a sense of purpose, and another study reports that a whopping 80% of Gen Z sought jobs that were aligned with their interests and values.[33] Compare this with Boomers, those former flower children who came of age in the 1960s and 1970s. Only 47% of Boomers felt values and interests had much to do with their work.

Another set of ideals that is important to VNs, linked to their radical acceptance of others, is expecting the working world to be accepting and respectful of them – as they are. Virtual Native TikToker @adminandeve posted a video explaining that she rescinded her application for a position earning a six-figure salary for two reasons: (1) The company would require her to dye her hair a natural color, and (2) the two interviewers on the Zoom call repeatedly muted her so that they could talk to each other. Her followers were shocked by the lack of respect indicated by both behaviors. One commented, "People always tell me to take my nose ring out for interviews – nope. I can't work for an organization that thinks that impacts my work."[34]

A fourth driver for job change is flexibility, particularly around commuting and working from home. McKinsey found that Gen Z workers are nearly 60% more willing to quit than older generations if the option to work remotely becomes unavailable. VNs are also more likely to answer job postings that specifically mention flexibility of work location.[35] Virtual Natives are technologically the most advanced generation alive, and they have no time for pretense or outmoded systems and protocols. They won't travel several hours every single day and wear a jacket and tie just because that's the norm. Things most generations passively accepted as "work culture" are now being questioned, because Virtual Natives now have the digital tools to do things in ways that make more sense to them.

In one recent study, VNs were split almost evenly between wanting to have the personal connections developed by being physically present in an office setting, and between wanting the freedom and convenience of working from home. As one VN put it, "We don't want to work less, we just want the flexibility that hasn't been associated with the typical work atmosphere of all worklife, and no balance. We want to work smarter, not harder," and that includes being able to choose where and when they make their contributions.[36]

As a corollary to the popularity of remote work, a study by Manpower Group found that "industry sectors which are either less able to offer remote work or have been slower to embrace it – including construction, finance, hospitality and manufacturing – have faced some of the biggest skills gaps for all types of job."[37] If it can't be done in a way that gives them at least some control, Virtual Natives just aren't excited about it.

It may be that this VN outlook of questioning the old ways, and only taking jobs that respect their personhood and ability to make their own decisions in life, will give them the last laugh in the end. "People who switch jobs more frequently early in their careers tend to have higher wages and incomes in their prime working years," says Henry Siu, a professor at the Vancouver School of Economics.[38] When the average tenure of a young person at a company is no longer five years, but more likely one or two, that's not a bad thing. According to Siu, "Young people aren't quitting more. They're experimenting more."

We're All Freelancers Now

One way in which Virtual Natives have been able to balance their need for an income with their desire to remain in control of their personal destiny is through freelancing. Spurred by the emergence of platforms including Upwork.com, Freelancer.com, and Fiverr.com, digital tools and internet availability have enabled freelancers to code or consult for Chicago clients from Bali. The burgeoning freelancing revolution is estimated to already be a $1.5 trillion economy, engaging as many as 500 million part- and full-time professionals through more than 1,000 platforms, on a worldwide basis.[39] Thus liberated from location and time zone, Virtual Natives who freelance are able to exercise control over when, where, with whom, and for which cause they will work, with the unpredictable variety that comes with freelancing providing the fresh learning that they crave.

MBO Partners shows that the total number of "independents," people who earn money periodically by working at least monthly as an independent, grew 69% from 2020 to 2022. The report further found that the proportion of gig workers "who see independent work as less risky than a traditional job" rose from 29% in 2021 to 33% in 2022.[40] Think about this for a moment: A *third* of gig workers find independent employment to be *less risky* than becoming a full-time employee. For VNs, with their high tolerance of failure and willingness to try something else if their first idea doesn't work, getting locked into a soul-crushing, freedom-killing job is a higher life risk than not having a steady income. This is a significant generational shift, indeed.

Whether they work as long-term freelancers or skip around from employer to employer at a frequent rate, Virtual Natives do have something that employers need: deeply lived insight into interconnected social worlds, changing consumer behaviors, and the rise of digital goods and services. In an era of increasing digitization and virtualization, these skills are now invaluable for corporate survival. Can anyone blame Virtual Natives for wanting to be recognized and adequately rewarded for their unique abilities, or begrudge them for wanting to exert control over their lives?

Smart organizations are leveraging the skills of Virtual Natives by purposefully soliciting their feedback on subjects of interest to them, using the tools that they already use. As excellent examples of this, both Riga, the capital of Latvia, and the City of London have initiated urban planning outreach projects with Virtual Natives in Minecraft. In Riga, city planners are using an interactive Minecraft map of the city to include schoolchildren in a dialogue about the future of the city and its inhabitants.[41] In London, Mayor Sadiq Khan appears in a Minecraft version of the suburb of Croydon, inviting students to redesign the town center as a "greener, safer, and more prosperous place for everyone" under the 2023 Design Future London Schools Challenge.[42] The point of inviting teenagers to contribute their thinking to City Hall, according to Khan, is "to inspire young Londoners to develop an interest in designing beautiful, affordable and sustainable places to live, work and visit." And Minecraft is the perfect medium to indicate to VNs that their style is understood – and welcome.

A job fair held in the Metaverse in Japan in January 2023 attracted over 2,000 attendees, who appreciated both being able to attend the fair remotely, and the anonymity granted by being represented by avatars, which allowed them to ask questions of their potential employers on delicate matters without feeling that they were jeopardizing their chances of being hired. Virtual Natives appreciate companies who meet them where they already are.[43]

In contrast, authors Renee Dudley and Daniel Golden recount in their 2022 book about the FBI's efforts to track down criminal hackers, *The Ransomware Hunting Team*, that the FBI was far too hidebound to accommodate the needs and preferences of Virtual Natives. Instead of recognizing that the extremely digitally savvy people that they needed to recruit might have a different profile from their traditional agents, the FBI retained its law enforcement–focused recruitment requirements for all new agents, even while hiring white-hat hackers. "The FBI wanted its cyber agents to be athletic college graduates with relevant job experience, who also had to be willing to shoot a gun and relocate their families," a bad fit with a cohort more likely to have spent long hours in front of a keyboard instead of pounding out squats at the gym. Those computer experts who did make it into the agency were looked down upon by older agents, who called them "dolphins," meaning that "they're intelligent and have their own language, but you can't understand anything that they say."[44]

Not surprisingly, with this mindset, the FBI has had difficulty attracting the people that they need, especially when highly skilled computing abilities are so very well compensated and welcomed in the private sector. The FBI's solution has been to partner with outside companies and informants. The problem with using a constant stream of outsiders is that the FBI never develops its own institutional learning, which is disastrous in the fast-moving field of cybercrime. And not all partners will put the public good ahead of their own profits.

And let's face it – Virtual Natives really are the ones who best know the digital underpinnings of our society, and what can be accomplished with them. As 23-year-old Connor Blakely, who runs Gen Z marketing house Youth Logic, points out, "The world is full of kids who are figuring out solutions to problems. . . Gen Z has a certain lens and a certain set of skills to be able to translate the future into something that is practical for consumption and inherently valuable." In fact, Blakely believes that the smartest companies are the ones with Gen Z representation at the very top. "I spend a lot of time in boardrooms [representing Gen Z insights] and I have not seen an example of us doing that for a company where they didn't thank us."[45]

Although this chapter has been focused on the workplace, the larger Virtual Native theme that this discussion reveals is that of the importance of agency and control. Encouraged by their childhood experiences of being rewarded for creating digital content and their other online activities, Virtual Natives have a strong sense of the value of their abilities, and will take action if they feel that their workplace does not give them latitude to make decisions for themselves. Examples abound of VNs making videos of their dramatic and triumphant quitting moments, such as Marina Shifrin's An Interpretive Dance for My Boss Set to Kanye West's "Gone," which immediately went viral, got her invited to appear on TV, and ended up with her getting a new, much better job.[46]

From quiet quitting to rapid job rotation to freelancing, Virtual Natives have options, and use them. VNs are one of the most decisive and empowered generations we've ever seen, and only the companies that speak their language and respect their terms – and their nose rings – will receive the full benefit of their considerable digital skills.

9

Send Pics

WHEN A VIRTUAL Native needs to learn about something, they're far more likely to seek out a video on the subject on YouTube than to look for a text-based answer on Google. VNs also prefer images over pure voice – via quick video chats – and for socializing, use other forms of communication that generally do not depend on the written word, and especially not on the handwritten word. This poses profound questions about what literacy is and what kinds of communication will retain both relevance and meaning for this generation going forward.

When Annika's son Finn was in the fifth grade, the biggest event of the year was Science Camp. The entire class got to go live in cabins in the Santa Cruz Mountains for a week with their teachers and nature guides, who would take them through the forest and show them the joys of finding banana slugs and cracking wintergreen Lifesavers into sparks with your teeth after dark. Practically the whole school year had built up to this enormous adventure which was for many children their first real length of time away from their families.

To give kids a reminder of home, the school suggested that parents write their children a letter and mail it to the camp ahead of time, so that it would arrive on about the second day of the adventure, just when kids were most likely to start feeling homesick.

"What a good idea!" thought Annika, and she sat down and wrote what she thought was a funny, yet reassuring card for Finn. As she mailed it, she thought warmly of the pleasure that getting the surprise card (the kids didn't know the parents were doing this) would give Finn, out there in the forest, surrounded by banana slugs.

The kids went to the camp, had a great time, and came home again. Finn was elated, full of stories about learning to handle nonpoisonous snakes and the nightly singing competitions around the campfire.

After a while, Annika asked, "Did you get the card I sent?"

"Oh, yeah, that," Finn replied offhandedly. "I couldn't read it."

". . . what do you mean, you couldn't read it?"

"Just what I said. I couldn't read it. You wrote in cursive!"

Annika's jaw dropped. "Are you telling me that you can't read cursive? Didn't they teach that already?"

"Well, yeah, they taught it to us in fourth grade. But we never read it or write it, we type everything. Except maybe our signatures. Outside of the classroom I've never seen cursive before the card you sent me."

First-grade teacher Claire, in her twenties, confirms the dominance of the typed, rather than handwritten, word among the Virtual Natives she teaches in Arkansas. "The kids in my class all started first grade *already knowing how to type*, from using computers and playing games at home. They use two or three fingers on each hand, which isn't exactly the official way to do it, but their typing is definitely far faster than their handwriting is, even at the age of six. I think I was the last generation who had to be taught typing in high school."

Fast-forward to 2020. It's the height of the pandemic, and as Annika passes the doorway to her daughter Sasha's room, she sees Sasha sitting at her desk, adding a soundtrack to a video that she has filmed herself.

"Hey, Sasha, what are you making?"

"Oh, this is my gym class assignment."

"Gym class? You have to make a video for gym class?"

"Well, while we're doing at-home schooling, it's hard for us all to exercise over Zoom. So the teacher gives us assignments that we have to do on our own time, like a certain number of jumping jacks, and we have to film ourselves doing it and hand it in for credit."

"That seems like a sensible solution. But doesn't it take your teacher a really long time to check the videos of all of the students in each class?"

"He's asked us to speed up our videos, so we can squish about 30 minutes of exercise into less than a minute."

"So why are you adding music?"

"Because I want my video to be really cool. Look, I've added visual effects, too – fireworks when I finish!"

Let's put these two scenarios together. On the one hand, Virtual Natives such as Finn struggle to read cursive. On the other hand, when given a video production task as an assignment, it's so easy to do, and comes so naturally, that Virtual Natives such as Sasha think nothing of adding in extra music and visual effects.

We're standing on the brink of a profound generational change in how language is used and perceived. Virtual Natives are much more likely to communicate with video, photographs, memes, and abbreviated text language than they are to write out a paragraph like this one that you're reading right now.

A Generational Shift

Neuroscientist Dr. Maryna Yudina of the Pacific Neuro Center does brain scans on people of a wide range of ages as part of her role as a performance enhancement trainer. In the course of her work, she has noticed significant and recurring differences between younger brains and older ones, while emphasizing that this is anecdotal evidence, not something she's specifically studied. Some differences are likely linked to age and experience ("younger brains are less organized, more chaotic"), but others seem to indicate cultural changes. Most notably, she sees that in general, younger brains have much more prominent development of the visual cortex, but much less development of fine motor control. "Kids today are using keyboards to write, which requires less fine motor control than using a pen does, and we can see that change in the brain," she relates. She often recommends that her clients take up something that requires fine motor control, such as playing an instrument, so that they can develop and maintain the brain-finger link that gives people the ability to do very delicate, intricate work with the fingers. According to Dr. Yudina's informal observations, the digital world reduces our need for physical refinement, while at the same time increasing our brains' abilities to perceive and process visual imagery.

And the visual imagery that Virtual Natives are spending their time looking at isn't printed text. A few years after Science Camp, Annika gave her son Finn, then 16, a copy of a book that she had loved in high school: *Einstein for Beginners*. Finn was incredulous that his mother was handing him, of all things, a book. "Come on, Mom, I'm not going to read a book."

Annika was ready for him. "Actually, it's a graphic novel."

Only then did Finn start to show interest in the gift. "Well, if it has a lot of pictures, then okay, maybe I'll look at it sometime."

We should point out at this juncture that Finn is a good student who takes Advanced Placement classes and does indeed read the books assigned to him in English class. It's just that those are the only books he ever reads – he, and many of his fellow Virtual Natives, don't read for fun, as their parents may have done.

What they do instead is watch videos, often the videos created by other Virtual Natives.

"There's a whole genre of videos on YouTube that are about showing you better ways to do everyday things," says Selina, mother of two Virtual Natives. Examples of what Selina has learned from her VN children include using an empty plastic water bottle to separate eggs (fast and effective), and that there's a little arrow on the gas gauge in your car that indicates which side of the car the gas tank is on. ("A lifetime of driving, and I never noticed it before!") These bits of information are often faster and easier to explain with a video example than with text.

It's not just YouTube videos where Virtual Natives find information about life. In a nod to the digital-is-real/physical-is-real relativity that we discussed in chapter 2, VNs also find that games and online simulations can be excellent teachers.

Actor Keke Palmer, in her twenties, was in her third trimester of her first pregnancy, and wanted to know more about what to expect during the actual process of childbirth. This is a subject about which many, many books have been written, but Keke was more interested in living through the experience digitally than just reading about it. She ended up in the game *The Sims 4*, which, with the addition of a few mods, was able to give her the full experience of pregnancy, starting with pregnancy symptoms and ultrasounds, and ending up in a range of possible childbirth options, including at-home births, Cesarean sections, and natural births in the hospital. All that's missing is the pain.

Keke tried them all. After mentioning on Instagram that she was using *The Sims* to explore her birthing options, fans asked her to start streaming her *Sims* play on Twitch. She taught herself how to do this, launched her Twitch account in a day, and had gained 30,000 subscribers only five days later.

Keke is thrilled with the experience that she's gotten by working through different versions of labor and birth in *The Sims*. "I would absolutely recommend that other people who are pregnant see what kind of skills they can develop by playing games and going into the Metaverse," she enthused.[1]

For VNs, interactive games and virtual worlds let them do far more than just imagine something after reading about it. Digital interactivity lets them virtually experience it, giving them a deeper understanding of the physical-world reality, and to learn about resources and options. No matter how accurate written text may be, a picture is, after all, worth a thousand words, and VNs feel that reading abstract words on a page is no match for experiencing a video representation of the same thing.

Virtual Search

Virtual Natives' preference for learning and communicating by video, rather than with the written word, extends to their usage of search functions. In 2022, Google Senior Vice President

Prabhakar Raghavan admitted that 18- to 24-year-olds don't use keywords to search for information. "In our studies, something like almost 40% of young people, when they're looking for a place for lunch, they don't go to Google Maps or Search," he said. "They go to TikTok or Instagram."[2] He also pointed out that, unlike older generations, Virtual Natives do not generally have any experience at all with paper-based forms of information, such as maps. For them, a two-dimensional map on a phone is not a familiar, interesting, or efficient way to convey information. This has made it imperative for Google to break away from legacy information formats and rethink how it can help people to position themselves in their environments. Augmented Reality cues integrated into the phone's camera, for example, are far more in line with VN expectations than a little blue dot on a two-dimensional grid.

Among Virtual Natives, hard copies are more of an annoyance ("What am I supposed to do with this piece of paper?") than an authority. This can cause conflict with older generations who still rely on paper-based systems to function.

As Marissa, a Millennial, relates, "Recently, the electricity went out and I didn't know who to call. So I went to the building's concierge and spoke with Jennie, who's around 20. She told me to report it on the app but my phone wasn't charged – because the electricity was out. Ugh, I was so frustrated! For future, I asked her, can you just print me a list of everyone to call? You know what Jennie did? She took a picture of the website, then texted it to me, which of course I couldn't access with my dead phone. All I wanted was a printed sheet of paper, but she just decided a picture was better."

Another problem posed by the Virtual Native shift away from classical text-based language, and toward audio and video sources, is – how do you search for content among audio or video files? If you're looking for information about how to boil an egg on YouTube, you will only find the videos that have been labeled with actual text along the lines of "boiling an egg," because there is currently no mechanism to search the content of videos out on the web for moments of egg boiling.

And this matters for digital platforms, including the nascent Metaverse. What does the search engine of the Metaverse look like? If you want to find out where Chipotle is giving away free burritos in the Metaverse, or locate your friend's avatar, how do you know where to look? Right now, you can only find these things if you already know which platform they're in, and then where to go within the platform. If you don't know that information beforehand, you'll never find what you're looking for.

In *Connections*, the glorious 1978 documentary and book by science historian James Burke, he points out that the real revolution that kickstarted the Renaissance wasn't the printing press, it was the index. After books had been printed, the knowledge within their pages was still effectively inaccessible until someone came up with a system for identifying how to find what you were looking for without having to read every page of every book on the shelf. Powerful search engines performed the same transformative miracle for the internet, and the search engine of the Metaverse – whatever it turns out to look like – will be rocket fuel for the development of those virtual worlds. It will very likely not be text-based.

Connecting Across the Generational Divide

Standing at the dawn of the Metaverse, we have a generation of Virtual Natives whose first language is not only digital, but image-based, rather than centered around written language. Like their 1990s counterparts before them, whose workflows developed around PCs and floppy disks, rather than the typewriters and carbon copies that preceded them, today's Virtual Natives have no reference to the workflows of the past.

Try asking a Virtual Native if they own a printer.

In all likelihood, they don't, and they'll probably laugh at you for even asking the question. Virtual Natives are more likely to bring a computer or phone to a meeting than a pad of paper. In fact, most Virtual Natives will never have folded a letter in three and posted it in the mail. They often have no idea even how to address an envelope, and they definitely don't know how to write a check. Paper is no longer essential for tasks; it is entirely a relic of past work habits.

Employers must therefore understand that their future employees will not think in terms of handbooks, distribution lists, or anything that stems from physical protocols or conventions. And Virtual Natives won't see a need for them. Instead, they'll be expecting to be onboarded through virtual reality or Zoom, the platforms and formats with which they're familiar.

Just as our friend Marissa demanded a list on paper and received a text in return, Virtual Natives will translate their understanding of a task into what makes sense for them. This will inevitably clash with older processes and procedures, but need not create friction. If embraced, it presents an opportunity to introduce much-needed innovation.

As we saw in the previous chapter, wise companies understand generational transitions and leverage them in their dealings with the Virtual Native contingent. Gen Z market research firm Knit knows very well that the Virtual Natives they survey find making videos far easier than typing out long written passages. As a result of this knowledge, Knit specifically asks its research subjects to create and upload videos when they answer free response questions, instead of asking them to fill out a text box. VNs have absolutely zero problems with recording and uploading a video of themselves stating their opinion – they do it all the time on TikTok and Instagram, after all. Knit then uses AI to parse the tens of thousands of videos that they receive in the course of their surveys and to distill the responses into usable data. Knit's combination of video and AI, and its ability to leverage the two to elicit the most natural responses from its Gen Z market research base, offer an intriguing hint about how we may increasingly communicate and find meaning in the future.

A world far less reliant on textual communications in daily life may be an unsettling vision for many. But Virtual Natives, with their deep, preexisting comfort with video and audio information sources as meaningful primary authorities, and a scant regard for the importance of actual written words, will step right into these digital worlds and feel at home immediately. They are the ones who will be most comfortable creating and consuming content in low-text information systems.

10

The First Cyborgs

When ChatGPT burst on the scene in December 2022, teachers around the globe went into a frenzy. How could they prevent their students from using it? How could they spot when a student had "cheated" and used it to write their essays? But Virtual Natives had an entirely different reaction. "Cool, this is just like having a calculator in my math class, except this is for English class. Grunt work, goodbye!"

Virtual Natives already use digital tools to do almost everything of importance in their lives, from their homework to staying in touch with friends, to entertaining themselves on a rainy afternoon, to generating eye-opening levels of income while still too young to vote. It is only natural that they view Artificial Intelligence (AI) in all its forms as a natural partner and collaborator, not as a threat. It's still early days, but all signs point to the likelihood of Virtual Natives being the generation that most intensely embraces AI, with potential world-changing consequences.

"Tell Us About the New Digital Natives in Virtual Spaces"

We consciously set out to publish the world's first book cover co-designed by AI, using Midjourney (Thanks Carlos! Thanks Wiley!); it wasn't long before we wondered how else we might use AI.

After all, writing a book takes a lot of research, thinking time, and especially physical typing time. And time is one resource that, once gone, can never be replaced.

So, after several intense months of research and writing, we started to wonder if we could get away with letting AI finish writing this book for us.

We opened ChatGPT and fed it prompts like, "Tell us about the habits and preferences of online communities in virtual worlds." We asked it to opine about Gen Z in relation to a wide variety of different subjects, and found that, in general,

ChatGPT wasn't able to connect disparate concepts, or project how converging trends might evolve. *Sigh*. So we kept writing.

But Ammaar Reshi, 28, was luckier.

A student of computer science in the UK, Reshi moved to Silicon Valley in his mid-20s for a job at a prestigious fintech firm, working as a design manager. When his best friend had a baby, rather than order a book for the child, he decided to write one himself, using ChatGPT for the text and Midjourney for the illustrations. He worked on the project over a single weekend, feeding prompts into the AI to develop the story and refine its details.

"Anyone can use these tools," Reshi said. "It's easily and readily accessible, and it's not hard to use, either."[1]

After only 72 hours, he uploaded the book to Amazon's Kindle Direct Publishing, and tweeted, "I published a children's book co-written and illustrated by AI!"[2] The book is about a girl named Alice and her Robot, Sparkle. Though we haven't read it, the cover is actually pretty adorable. But other children's authors and illustrators were enraged by Reshi's audacity, expressing their concerns about both the ethics of AI-assisted writing as well as the potential, children's author Josie Dom said, of "a proliferation of poor-quality stories, both on the writing and the illustration side."[3]

It hardly needs pointing out that most of these authors complaining about Reshi's foray into AI-partnered authorship are a lot older than he is. The reaction of Virtual Natives is much more likely to be like that of Mayling, age 15. When asked about the ethics or appropriateness of using AI to generate text for a work that you would then sell as your own, she shrugs. "Why not? I mean, these good tools are already here, so I don't see a problem with using them." Her brother Zihao, 17, is blunter. "Dinosaurs don't survive, man. Evolve or die."

Clearly, we are not the only ones testing the use cases of AI. Students around the world are using it for a range of applications from writing papers to helping them perfect their code. YouTube videos like "How to Use ChatGPT to Write an Essay or Article" and "How to Use AI as a Partner in Your Research" abound. As Generative AI of many types gets baked into our daily work tools, like Google and Microsoft Office, how will that affect learning and the future of work itself? Surely those who have the most experience with partnering with these powerful tools will be the ones who advance the furthest, the fastest.

When philosophy Professor Antony Aumann at Northern Michigan University sat down to grade a set of papers on world religions, one in particular stood out. Thorough and well structured, it was rigorously argued, and well supported with pertinent examples. The paper was easily the best in the class. The only catch was that it was written by a student who had not, until that point, ever delivered work of that quality. Suspicious, Aumann confronted the student, who confessed to using ChatGPT to write the paper for him.[4]

From poetry to literary analysis to screenwriting, the premise of ChatGPT is that it can take a text prompt, such as, "Tell me a story about a girl with a magical black cat that solves mysteries" or "Describe the symbolism of nature in Shakespeare's *Macbeth*," and it will develop an entire essay, plot, or even screenplay in seconds. Children today will grow up with these tools as part of their standard toolkit. What does that portend for the future of learning, of creativity and work itself?

For Professor Aumann, the lesson he learned was clear: He would have to restructure the way he teaches. Whether through oral arguments, or in-class essays, or changing the nature of the composition topics that they set, teachers everywhere are learning how and what to teach in the age of AI.

But trying to freeze out all use of AI support is misguided. Perhaps a better approach is that of high school English teacher Kelly Gibson. Located in a rural, low-income region in Oregon, she quickly realized that the digital divide between those of her students who have internet access at home, and those who don't, would be driven even wider if only those with internet access were able to play around with ChatGPT and develop their AI skills. She therefore resolved to bring ChatGPT into her classroom, so that she could teach AI-assisted critical thinking to her students. She's currently experimenting with having students write original thesis statements about the works they're reading (*Death of a Salesman* is up first), feeding those into Chat-GPT to generate essays, then critically assessing the results and using their own ideas to improve on what the AI delivers to them.[5] It is a savvy English teacher indeed who realizes that critical thinking, good communication, *and* AI management skills are all necessary parts of a modern education.

Machines Are for Grinding

Aneesh Dhawan, the CEO of the Gen Z market research company Knit whom we met in the last chapter, has built his entire company around embracing AI. Not only did his team develop an AI tool that allows Knit to derive meaning from the video responses of their test subjects, as described earlier, but Knit also uses AI to help its clients generate the right kind of survey questions to help them get the results they want.

If you've ever tried to write survey questions, you quickly learn that it's trickier than it looks to write a question that gets the response you're looking for without inadvertently warping the results with some kind of unconscious bias. Dhawan realized early on that Knit could use Machine Learning to create an AI tool that would help his clients, who were not research professionals, generate their own well-crafted survey questions at a fraction of the cost that it takes to hire an experienced human to do the same job. And like all Machine Learning, it improves as more and more data is fed into it over time. They're a relatively small company, but using AI in both the formation and analysis of their surveys lets them accomplish far more than they could if they were just doing everything by hand. "And the work that the AI does is all the tedious stuff," he notes. "We still get the fun of looking at the results and drawing conclusions."

Because we're dealing with Virtual Natives, who are masters of monetizing their digital skills, marketplaces have already arisen for those who have developed expertise in getting Generative AI to give the results that they want. PromptBase is one such marketplace, in which artists post the results of different prompts in Dall-E, Stable Diffusion, Midjourney, and GPT3. Buyers who are looking for that kind of visual style can pay a couple of dollars to the creator and learn exactly what words that they used, then use the prompt to develop their own creation in that mode. The range of styles, effects, and imagination on display in the PromptBase marketplace is awe-inspiring, and demonstrates that even though Generative AI tools are accessible to all, it still takes a human to have the original thoughts that generated this incredible array of art. It's the human and the machine working together that gets the best results, not either one on their own – which is what Virtual Natives already know.

Learning, Development, and Artificial Intelligence

Let's take a moment to explore the promises and downsides of AI for learning, and how both students and teachers are testing the edges to incorporate or exclude AI in education. But first, what is it, and why is it such a big topic today?

Generative AI holds great promise for the future of creativity. In addition to generating text-to-speech or text-to-image results in tools such as ChatGPT or Midjourney, it can also convert text to 3D objects for game spaces, create original music, and more. It has the power to transform what we hold in our imaginations into something we can visualize and share with others, without needing to know how to code. It can even pretend to be us!

Microsoft has released a neural codec language text-to-speech model called VALL-E that can mimic voices and convert them to complete sentences, based on only about three seconds of the speech of the person they are emulating. The program can generate conversations using your voice and speech patterns accurately enough to fool your friends, at least over the phone.

Nicholas Carr, author of *The Shallows: What the Internet Is Doing to Our Brains*, contends that while the internet has given us powerful tools for finding information and connecting with people, it has also made us into "lab rats constantly pressing levers to get tiny pellets of social or intellectual nourishment."[6] In his book, he suggests that 3D tools and, by extension, Artificial Intelligence, expand new strengths in visual-spatial intelligence while at the same time undermining "mindful knowledge acquisition, inductive analysis, critical thinking, imagination, and reflection."[7]

Will the collective ability of future generations to concentrate, absorb, and analyze information truly be affected by automatically generated media? For Carr, surfing the internet is a reductive act, reversing human intellectual development. As he puts it, "We are evolving from being cultivators of personal knowledge to being hunters and gatherers in the electronic data forest."

UCLA developmental psychologist Patricia Greenfield, a longtime observer of the relationship between children and technology, disagrees. She maintains that, "We're assuming technology turns the brains to mush. Every experience we have affects our brains. There's nothing special about using technology that affects our brains. Reading affects our brains. Everything we do strengthens some neural connections, possibly at the expense of others." Even the printed word, she has argued, is a form of technology.[8]

Greenfield was an early observer of the long-standing tensions between technology and formative education. In the pre-internet Pac-Man era, she was one of the first to suggest that video games could enhance the development of complex skills like spatial visualization and inductive reasoning.[9] By jumping into a game, learning its modalities and devising a strategy to win by iteration, children repeatedly construct and test hypotheses.

"If I do this, then what happens?" This approach is, in fact, the essence of the scientific method. So is it truly harmful and reductive for children to interact with powerful new computing tools, or are new generations merely entering another phase with new and inventive ways to solve problems?

It's a new world. Goodbye homework!
—*Elon Musk tweet, January 4, 2023*[10]

Can We Beat the Bot?

English teacher Kelly Gibson is only one of the educators who is putting thought into how to bring Generative AI into the classroom as an integrated part of her students' learning experience. Ironically, when she uses ChatGPT to help her students improve their *Death of a Salesman* essays, she has to bring her personal computer into the classroom, since her school district has banned the use of ChatGPT-like tools on the school internet.[11] The implication of her school district's move is, of course, that they're afraid that the students will use ChatGPT to cheat.

Stefan Popenici, an education and leadership specialist at Charles Darwin University in Australia, points out that technologies such as pocket calculators and spell checkers, which are today deeply integrated into our computers and phones, were hotly contested and controversial within the education community when they were first introduced. Similarly, he expects AI to overcome opposition by being genuinely useful over time in the classroom, and even envisions a future in which resource-strapped universities include highly capable "teacherbots" in addition to human professors.[12]

Examples of how we might embrace, rather than fight, ChatGPT perhaps include giving students a ChatGPT-created essay and asking them to fact-check it, subtly teaching them that they can't just blindly trust passages produced by Generative AI tools without question. Or for a research assistant to use it to find resources, or for a teacher's assistant to use ChatGPT to review an essay, code block, or mathematical formula and to provide suggestions on how to improve their work. Educators should certainly teach prompt crafting, much as they already teach their students how to use TI-84 graphing calculators in math classes today, having eventually overcome the doubts expressed by many teachers when pocket calculators first came out.

One standout example of how to use AI in academia came from Edward Tian, a 22-year-old student at Princeton. As a junior majoring in computer science and minoring in journalism, he is well suited to address the challenge of our times: spotting ChatGPT in what are meant to be human-only essays.[13]

He spent his January 2023 winter holiday building GPTZero, "an app that can quickly and efficiently detect whether an essay is ChatGPT- or human-written." His Tweet announcing his creation went viral, amassing over 7 million views, and within a week of its launch, over 30,000 people had tested GPTZero. While Tian doesn't believe ChatGPT should be banned from schools because of its enormous potential for usefulness, he does believe in transparency.

"It doesn't make sense that we go into that future blindly," Tian reasons. "Instead, you need to build the safeguards to enter that future."

We're All Centaurs Now

In the world of competitive chess, "freestyle" chess is a format in which humans can consult outside sources before they make their move. Matches are timed, so the time for this consultation is quite limited. Within these restrictions, a new kind of player has emerged, known as a "centaur."

Instead of the half-horse, half-human creatures of myth, chess centaurs are humans playing the game while allied with a source of Artificial Intelligence. "A centaur chess player is one who plays the game by marrying human intuition, creativity and empathy with a computer's brute-force ability to remember and calculate a staggering number of chess moves, countermoves and outcomes."[14] The computer can make suggestions, but it is the human who ultimately selects and makes the move.

This combination of human and computer turns out to be the most powerful alliance the chess world has seen. A good centaur can defeat a great grandmaster, one who would readily win victory over the human half of the centaur if they were playing on their own.

The concept of the centaur is the perfect metaphor for summing up the relationship that Virtual Natives have with computing in general, and with AI in particular. VNs know from their lived experience that they can do far more by working with computers than by ignoring or denying their capabilities. They will be the first ones out of the gate as new AI capabilities develop, founding disruptive new businesses and finding insightful ways to improve their own lives.

For those of us in other generations, the takeaway is that if we're not careful, VNs could use their AI expertise to leapfrog us. As noted tech advisor Shelly Palmer explains, "There is a profound difference between losing your job to a machine and losing your job to another person who has developed [Generative AI] skills you have chosen not to develop."[15]

Virtual Natives are the ones developing those skills. The centaurs are already here.

PART

2

The Revolution Starts Here

THESE, THEN, ARE the characteristics that set Virtual Natives apart from other generations and mindsets: VNs use digital tools to live life on their own terms; they view the physical and digital worlds with equal legitimacy; they are widely accepting of identity fluidity in themselves and others; they anchor their view of authority in the twin attributes of charisma and authenticity; they are deeply expert in digital ways; they know how to make money from their digital enterprises; they are used to calling their own shots and can react badly when their ability to find their own happiness is limited; they communicate in images rather than text; and they are already leaning hard into the vast possibilities opened up by collaborating with Artificial Intelligence.

Now that we're familiar with the motivations and behaviors that Virtual Natives have developed while still largely teenagers, let's explore how they interact with the wider world and extant systems in it. This is particularly important to do now, right as Virtual Natives on the brink of adulthood begin to leave their childhood homes in large numbers to enter the adult worlds of work, bureaucracy, and finance, all areas that they have had scant interaction with in the past. And this is where it gets interesting.

Virtual Natives are equipped for action unlike any generation that has preceded them and are already taking education, work, and their future stability into their own hands. How will they apply the digital tools they wield with such alacrity to traditional processes and systems, where "traditional" tends to mean slow, unresponsive, and self-serving? The creative mindset and tools they command are poised to upend many administrative processes that have been carefully crafted over time, setting VNs up to become powerful challengers of existing institutional cultures and assumptions in almost every area.

Potently aware of their own market value, and independent actors whose main focus is self-advancement, Virtual Natives are revolutionizing the future of work, education, and culture itself. In the following pages, we'll explain how it has already begun.

Strap in!

11

Meet the Degens

As WE HAVE seen, Virtual Natives often begin earning and controlling their own money online as teenagers, as soon as they're old enough to start streaming on Twitch or marketing their digital sneakers on Roblox. This makes them one of the most financially savvy generations ever. Couple this with the upheavals in traditional markets that have shaped their world-view, and the rise of alternative Blockchain-based methods for capturing value, and we see VNs unafraid to take risks that their elders may be more likely to shy away from. Looking ahead to the technologies that will drive new forms of exchange, we explore what ignites the imagination of Virtual Natives as they search for financial autonomy, and importantly, control.

The Fever Is Rising

The scene was ArtBasel, December 2022. The romantic lull of the tropical evening – palm trees adorned with tiny lights, like shining pearls piercing the neon-blue night sky, swaying gently overhead – contrasted sharply with the focused agenda. The talk was about parties, but these parties meant business.

"Where did you go today?"

"I went to Bit Basel and Gateway; you?"

"Yeah, I went to see Gary Vee talk. There was a private reception after so I did that, and then I swung by Nolcha. I'm hitting up Moonpay later – you?"

"Yeah, that's cool. I wanna go, but I hear there's a thing with Snoop on a boat. I'm trying to get the link for the QR code now. We should be back in time for Moonpay or the Degen DJ fest later. I mean, the night is young!"

"You in?"

Ever since Bitcoin began its meteoric rise in 2017, the heat of the new digital gold galvanized cryptocurrency acolytes from around the world who gathered in online chat forums such as Clubhouse, Reddit, TikTok, Twitter, and Instagram to monitor the currencies' progress. Rabid speculation surrounding Bitcoin and other altcoins (for "alternative coins") began.

To the *noobs* (the uninitiated, from "newbie"), a lot of the jargon that surrounds this space can be intimidating, as is the fervor with which people are pursuing it. In addition to the online chats, there are parties, meetups, NFT galleries, infinite opportunities to engage. Everyone has something to pitch or sell, or the insider scoop on a collection or currency they're hyped up about.

What is attracting young people to this new sphere? Is it pure speculation? What do they hope to gain, if anything, by participating in these conferences and meetups?

First, a definition would probably be in order. The term "Degen," shorthand for "degenerate," can be traced back to sports betting. Originally an insult, the term refers to gamblers who bet large amounts of cash without much knowledge or experience, and who are therefore likely to lose their bets. Today within the Web3 community, the term "Degen" is worn as a badge of honor. It is an invisible signifier, an indicator that you are not an observer but a member of the community; you live this. Being a Degen no longer means that you're ignorant, either. Today, new initiates to the space or community, the newbs, are actively encouraged to "DYOR," or "Do Your Own Research." This is easy to say, but harder to follow when the whole game seems to be around, well, following.

For some, being a Degen and participating in the Web3 world means trading alternative digital currencies (altcoins). For others, it's about collecting rare digital assets and collectibles, minted as NFTs. For others, it's a chance to get in on the ground floor of a company whose product or service will soon be invaluable.

In other words, it's many things for many people, but for everyone you encounter in these often packed venues in cities across the world, like a chance at winning Willy Wonka's golden lottery ticket, it's a dream worth chasing. And it's not just for the Silicon Valley elite. Poet W. H. Auden captured this sentiment in his libretto for the operetta *Paul Bunyan*, a story about man's conquest of nature:

> But once in a while the odd thing happens, / Once in a while the dream comes true, / And the whole pattern of life is altered, / Once in a while the moon turns blue.
>
> —*W. H. Auden*[1]

The NFT and Web3 events and conferences we have visited in New York, Paris, Miami, Los Angeles, Austin, and San Jose had their fair share of tech bros present, but as often as not, it's a colorful cross-section of people who are fairly hard to categorize. And maybe that's the point.

"It's wildly diverse," explains Felice Schimmel, a PR agent who represents several start-ups and is a passionate Degen herself. "You'd be really surprised how different it is from the typical Silicon Valley kind of tech circles. And that's why it's so awesome; it's open to everyone, and everyone is welcome."

Schimmel describes her early forays into the space, after Bitcoin took off in 2017 and she decided to learn more about altcoins. "I started to hang out in this Clubhouse group, and it was very diverse, there were of a lot of African American guys, super-knowledgeable. The whole group was about knowledge transfer, and they were totally welcoming, everyone was there to learn. For younger people, this stuff is pretty intuitive."

She's right. Virtual Natives are used to doing everything on their phones. Book tickets for a concert? Done! Replace your driver's license? Just a click or two away! Need a copy of your college transcript? Done! For VNs, opening a crypto wallet and collecting digital assets, or transferring money from a bank and converting it from USD into crypto with a few clicks, is a relative no-brainer. This is often a contrast with older generations, who may be fazed by the seeming impenetrability of managing crypto exchanges, wallets, assets, and ever-changing valuations.

Since the motives for participating in these crypto-focused events are varied, and the attendees themselves highly diverse, it seems the only commonality is the dream itself. What is truly connecting these people, what binds them as a community? The answer is both surprising and simple. And not very original. It fundamentally boils down to human nature.

Searching for Gold in El Dorado

In the northern Andes of what is now Colombia lies an ancient lake, called Lake Guatavita, meaning "most high," which was inhabited by the Muisca people between 600 and 1600 AD. It was at this high-altitude lake, the one that brought the earth and the seas closest to the sky, that the kings of the Muisca (or Chibcha) people came to be crowned.

According to ancient Muisca legend, King Sua was married to a beautiful princess from another tribe. Sua was very fond of beer and bacchanals. Eventually his wife, mother to his daughter, fell in love with another warrior. When King Sua discovered the lovers, he had the other man brutally tortured. "If you want his heart so badly," Sua hissed to his queen, "then you shall have it." He demanded that his servants remove the warrior's heart, and serve it to his wife. The anguished queen took her daughter into her arms and fled, plunging them both into Lake Guatavita.

King Sua ordered the priests to recover his daughter. They brought him the child, but upon taking her into his arms Sua saw she had no eyes. King Sua despaired. He returned the child to the water and decreed that from that day on, the best emeralds and gold filigree were to be thrown into the lagoon to beseech the gods for peace and prosperity for the Muisca people.

This is the origin story of the Muisca coronation ritual, during which a new king was covered in gold dust before leaping into Lake Guatavita, while his servants poured precious stones and gold into the water.

Spanish conquistadors took note of this legend and passed it on. As the illiterate sailors that made up their ranks passed the legend along by word of mouth, the legend grew. Soon, "El Dorado," or "the golden one," was no longer a story about a king, but a rich city made of gold, with enough wealth for a thousand men to become rich forever. During the period between 1530 and 1650, thousands of men made dozens of forays into the unmapped interior of South America in search of the city of gold.[2]

The search for gold in the Americas has evolved with time, and technology, and continues to this day. Each generation has inherited a slightly different legend. In North America, some sought it in the California hills, others in the railroads or commodities markets in Chicago, others in the oilfields of Texas or on the trading floors of Wall Street. In the past two decades, we've seen hedge funds sprout into billions, and dot-com college dropouts turn garage hobbies into powerful publicly traded companies. The American Dream lives in the heart of every hustler in every coworking space across the country. It's a start-up lottery, and it's there to be won, with dry powder, determination – and no small amount of luck.

Today's Degens see El Dorado in the $69 million paid for an NFT of a Beeple painting. They see it in the potential to decentralize the grasping finance and banking system, which makes customers pay a fee for moving, trading, or converting *their own money*. They see an opportunity to lend directly to other peers, like maybe to a musician, in exchange for a percentage of future revenues, or to collectively buy a valuable asset, like the LA Lakers, or an original copy of the US Constitution, or the *Mona Lisa*. In short, Degens see the potential of Web3 to radically disrupt industries and money systems as they exist today and make them faster, more efficient, and more responsive to the market. They smell blood and they want a part of it.

VNs are playing at the edges of a risky, decentralized, largely unregulated market that is open to anyone with a phone, a debit card, and a crypto wallet. They see an open field, ready for their arrival, just as the Great Plains were once open to early settlers keen to cultivate and tame new land and call it their own.

The innovations that Web3 enthusiasts imagine are not a fever dream. Much of it is grounded in the possibility for new technologies and currencies to create faster, more efficient prototypes, trading platforms, digital assets, services, and solutions than those that exist today. Virtual Natives are fully aware of the gaps left by outmoded customer experiences, and the shortfalls that exist between the customer promise and actual delivery. Existing business and banking protocols, services, and user-generated content remuneration models are staid, distorted, and ripe for disruption. VNs see the gaps between their expectations and today's business reality as "empty land," ripe and ready for innovation. Degens want to create the world as it should be, not as it is. Mainly, it's about being in charge of their own destiny.

Comments Schimmel, "I think most people are trying to gain control of their own life choices, the privacy of their data, and a say in where their money goes and how."

VNs are particularly wary of Web 2.0 because they were raised with social media like Snapchat, Facebook, and Instagram, and there experienced the full spectrum of emotions that participating could elicit, including joy, FOMO, desire, and cruelty. They are aware, perhaps more than any other generation, how detrimental exposing themselves constantly online can be for mental health. And that's mainly due to algorithms that drive clicks, and hence, revenues, which can end up promoting content that is contentious, controversial, or just plain nasty.

With Web3, things promise to be different. Not only will creators have greater influence and earning potential, they also hope to have more control. In the ideal Web3 world, users will no longer be required to give up their data or a large portion of their revenues in exchange for access to the internet through platforms like Google, Apple, or Facebook. With Web3 platforms, content creators can be both participants and owners; they can build and govern sections of new platforms where the consumer is at the center of the business model. As Schimmel explains, "It's a way not to be a pawn in someone else's plan."

To be sure, it won't happen overnight. The existing banking system and government in the United States, headed by a group of octogenarians who likely struggle with smartphones, is in no hurry to change the status quo. Everything's always worked this way, so what's the hurry to upend the system with technology? Especially if it means power might somehow shift from them to the youngsters on computers. To those who enforce the existing rules, this gold rush to rebuild the internet represents one of many they have seen in their lives.

In the poem "Eldorado," Edgar Allen Poe uses the metaphor of a knight seeking the lost city as a dismissive commentary on fruitless, fantastic dreams. Similarly, many of today's older and more traditional leaders see the Web3 gold rush as another wave of water hitting a hot beach. Soon enough, they expect it will recede or evaporate and another will eventually come and take its place. Either way, the beach remains unchanged.

Gaily bedight,
A gallant knight,
In sunshine and in shadow,
Had journeyed long,
Singing a song,
In search of Eldorado.

But he grew old—
This knight so bold—
And o'er his heart a shadow—
Fell, as he found
No spot of ground
That looked like Eldorado.

And, as his strength
Failed him at length,
He met a pilgrim shadow—
"Shadow," said he,
"Where can it be—
This land of Eldorado?"

"Over the Mountains
Of the Moon,
Down the Valley of the Shadow,
Ride, boldly ride,"
The shade replied,—
"If you seek for Eldorado!"

—Eldorado
Edgar Allan Poe

Which is more foolish – following a dream? Or not?

The passion with which Virtual Natives have entered the Degen cohort and explored the power of Blockchain and Web3 as part of their financial and investing strategies[3] is yet another example of how Virtual Natives eagerly use the new digital tools available to them to refashion existing systems to fall more in line with their own needs. They saw their parents and older siblings burned by the tech industry in the early 2000s, then a too-powerful financial industry in 2008, and are not interested in letting the same thing happen to them, so they are forging their own path, taking their teenaged earnings in Roblox and Fortnite with them as they go. Navigating the intimidating thicket of Web3 terminology and crypto platforms is a breeze for them, and for all of them, not just those whose gender and/or skin tone matches the gender and skin tone that dominate Wall Street.

Every generation has its El Dorado dream. For many Virtual Natives, that dream is embodied by the financial freedom represented by Web3. This approach may be easy to dismiss as the crypto sector sails into choppy waters, but just as the dotcom bust of 2001 was not the end of the internet, what will happen as these VNs graduate from college and begin to create their own tech and fintech start-ups? What happens when they become the ones making the rules?

12

You're Not the Boss of Me

Virtual Natives grew up with social media. Unlike Millennials, who were guinea pigs experimented on by app makers who sought to influence and manipulate behaviors, Virtual Natives are highly aware, digitally literate, and have always been surrounded by the ongoing cultural discussion about just how coercive digital media can be. They get the majority of their information from digital platforms, viewing content produced by their friends, peers, and other third parties. Overall, VNs are "more immune to the lure of misinformation because younger people apply more context, nuance and skepticism to their online information consumption," according to a review of multiple surveys by Axios.[1] In fact, they're so confident about their own ability to sniff out the inauthentic and scammy that they lampoon older generations' ability to do the same by creating their own misinformation – then sit back and laugh when the oldies take the bait.

The VN resistance to manipulation is also a sign of their desire for personal agency and control, rather than wanting to let themselves be controlled by others. Their awareness of how coercive media can be may explain their insistence on authenticity and integrity before they grant authority to any entity. The world as it exists is full of points of coercion and behavioral cues. VNs not only see this; they will expect things to change this as they push back.

The Airline Puzzle

On a sunny day on June 11, 2021, over two million people cleared US airport security checkpoints in a single day. The number is significant because it was more than double the volume of people traveling during the height of the pandemic in April 2020, and 75% of the volume recorded on the same day in 2019,[2] leading some to name the post-pandemic phenomenon "revenge travel."[3]

147

What's interesting is not the amount of travel itself, but the number of accounts of "air rage," or poor passenger behavior, post-pandemic. Verbal abuse and all-out brawls more than tripled globally during the pandemic in the United States, while rising across Europe as well. In fact, in 2021 alone there were nearly 6,000 cases of air rage on US airlines, far surpassing the previous year's total of only about 100–150 cases.[4] Much of this behavior was dismissed as pent-up aggression and superstition over mask mandates. Others blamed alcohol. But what if it turned out that the reason was something entirely different?

You may have noticed that planes are often cold when you board. Then they seem to warm up later, especially after any food has been served, around the time that flight attendants dim the lights. This sequence of events is not linked to, say, the plane naturally warming up with a lot of people aboard, but rather, is a conscious progression designed to control passenger behavior.

"When the flight attendants want to take a break, they call me to raise the temperature. They call the cockpit and say, 'If you could turn the temperature up, we could get some of these people to sleep.'" revealed Andy D. Yates, a former United Airlines pilot.[5]

Small increases of temperature during flights of five hours or more are commonplace. They create what is called "the Everest effect," which occurs when you are at high altitude with the same amount of oxygen as down below, but with lower pressure. The result is that you sleep more easily. You may also get more dehydrated, sluggish, and may experience worse jet lag. But at least you got some sleep.

While that may not seem particularly pernicious, consider another common airline phenomenon: front-loading. Why do planes always start by loading the people at the very front first? Forcing passengers to push past other passengers while struggling to load heavy bags overhead and bumping into those already seated seems pointless and, intuitively, it simply seems illogical.

Surprisingly, the real answer is that random seating, or "WAC" (Window, Aisle, Corridor), is the most efficient method of boarding passengers onto a plane.[6] In practice, however, the driver is economic, and aimed at giving greater value to those who pay more. To make a differentiation that drives the willingness to pay more for a ticket, the airline must deliver perceived value, or perks, to those who pay more to travel. For business class passengers, it seems that the greatest advantage to sitting in the front of the cabin is a question of efficiency given by the ability to board and disembark first, without having to wait. But there are other more emotionally driven motivations that have everything to do with this—like schadenfreude.

"Schadenfreude" is a uniquely German word that evocatively means "the joy of other people's suffering." Social psychologists suggest that for some, the joy of watching others push past you to get to their economy seats, while being seen to sit in comfortable first-class banquettes, sipping champagne, is one of the main rewards for paying extra. The quiet satisfaction of having others witness your comfortable life, spying at you enviously out of the corner of their eye, is part of the perceived value of the extra fee.

This friction is baked into the consumer experience. Entire systems have been created over decades to load passengers onto planes and manage their onboard experience. The control is entirely in the hands of the airline and its staff. Creating friction as part of the business model is, in this case, intentional.

Increasingly, customers want to take back control of their online experiences, and that is especially true of Virtual Natives. Companies that design friction into their service models with unnecessary customer complications or difficult user journeys will have difficulties catering to new generations who expect to control how they access content, data, or services. Along with personal agency, VNs value authenticity, the acceptance of others, and the closely related topic of social justice. Putting these all together, Virtual Natives rank ethics as one of their most highly prized values.[7]

During times of crisis, Virtual Natives expect companies to "step up" and contribute to society, leading by example. Nike supported Colin Kaepernick's decision to kneel during the national anthem by subsequently making him the face of their 30th anniversary "Just Do It" campaign, and was widely hailed for its ethics and praised by Gen Z as a result.[8] In contrast, a Pepsi ad that supposedly showed Kendall Jenner calming a tense moment between protesters and police was met with widespread condemnation for its unwelcome appropriation and trivialization of a serious social movement.[9] VNs can smell the difference between supporting a social movement and exploiting it, and they will punish those who fall onto the inauthentic/manipulation side of the line.

Which brings us back to air rage.

A paper published by the National Academy of Sciences has found that while alcohol, crowdedness, and flight time do contribute to incidents of air rage, the central reason is actually class inequality.[10] The authors found that incidents of air rage were *four times* more likely when coach passengers had to pass through a first-class compartment to get to their seats. "Evidence of class differences" was cited as the single most likely predictor of air rage, more than any other contributing factor. Others have suggested that the decreased quality of the experience in the economy section, like seats and legroom that continue to shrink over time, is another factor.[11]

It's not hard to imagine how some might feel indignant at having to struggle to hold their heavy baggage aloft – to save baggage fees – while hovering over someone in first class calmly sipping champagne, before squeezing into a seat with no legroom; it's an emotional Molotov cocktail waiting to explode. And sometimes it does.

We may not understand consciously that airlines are manipulating us by playing on our schadenfreude and envy to get us to pay for business class or "economy plus" seats to get sufficient space to accommodate our own legs, but it's happening. From the moment we book our tickets and are prompted to add food, baggage, and legroom as "upgrades," to boarding to get to our seats, airlines are playing on our desires for status and comfort to increase their profits.

And of course they're not the only industry that consciously whittles away the components of a positive customer experience in the name of shareholder returns; we only focus on them here because they're so visible. In another example from the world of consumer products, companies hide rising costs with "shrinkflation," or subtly reducing the size of their products so you won't notice that you're now getting less than you used to, for the same price you used to pay.[12] Profits win, customers lose. The high service fees charged by banks for moving your own money that we mentioned in the previous chapter also fall into this category.

As we've seen, the desire for personal agency and control is a recurring theme of importance for Virtual Natives. Theirs is a generation that lost a lot of it growing up, due to the combined impact of both economics and the pandemic. This desire is likely to manifest itself in the future in ways we cannot predict today, but what we have seen from them so far is that their sense of a brand's ethics, authenticity, and lack of manipulation determine their relationship with that brand. Authenticity and respect for the customer are likely to decree whether VNs cheer or condemn a show, celebrity, or a brand experience, and influence how they reinterpret that brand in immersive worlds. Some 40% of VNs have said they will ghost a brand that does not align with their values.[13]

The Virtual Native Megaphone

When it comes to ethics and corporate values, Virtual Natives don't easily trust companies. They grew up using social media and witnessed the tricks those platforms used to manipulate user behavior. But they also know how the game is played and can turn it against companies that they perceive as taking advantage of them.

Sierra, 20, is a case in point. When her grandfather died, his pension was supposed to be transferred to her grandmother. There was no legal problem, but months began to pass, and Sierra's grandmother still hadn't started getting the monthly money she was due. The pension company kept assuring the family that the case was being processed, but there had been a few unfortunate hiccups in the paperwork that had slowed everything down. Out of cash, Sierra's grandmother had to begin dipping into her savings to pay for groceries.

"We thought that our situation was unusual somehow, and Grandma was just unlucky," Sierra said. "But then my mother was talking to a lawyer who works with the elderly, and the lawyer told her that pension companies routinely take six months or more to start paying pensions; the lawyer had seen some of their clients come up against real hardship as a result." Sierra was incensed. "Look, I'm no econ major, but even I can see that delaying six months before starting to pay out the pension is six months that the benefits company is getting interest on that money, not my Grandma. I totally lost it when I realized it seemed to be just kind of standard practice to basically steal money from grieving widows."

Sierra's grandmother and mother were frustrated, and didn't know what to do other than keep calling the benefits company and hope that the paperwork logjam would move one day. But Sierra had different tools in her arsenal.

"I started rage tweeting about how the benefits company was ripping off my grandma for their own advantage. I made sure to include the hashtags that the company uses in their own advertising, so my anger would show up in the feeds of anyone following that company. I posted on as many platforms as I could think of. I told companies that were looking for a company to manage the retirement benefits of their employees not to pick this one, because they were predatory and didn't pay people what was rightfully theirs. Everything I said was true, based on my family's actual experience, and I absolutely didn't hold anything back."

It worked. Within three days, the benefits company had called her grandmother to tell her that the paperwork had been resolved, and she began receiving her pension within the week.

This is just one early sign of how Virtual Natives are likely to react as they grow up and move further and further into a world of systems and processes that were built before their clients were digitally competent. Companies whose business models, intentionally or not, are built around sheeplike behavior from their customers will find themselves named and shamed by a generation that knows how to use digital megaphones, and will use them indeed if they don't get the personalization and responsiveness that they feel they deserve.

The Age of Conspiracy

Several decades ago, all the birds in the United States were killed by the CIA and replaced with spy drones in an operation called "Very Large Bird."

"Wake up America! Birds are not real, they're a myth, they're an illusion!" shouted Peter McIndoe, the Gen Z college dropout and 2017 founder of the Birds Aren't Real Movement. (If you're wondering whether he was serious or not, he wasn't.)[14]

McIndoe explains, "It's taking this concept of misinformation and almost building a little safe space to come together within it and laugh at it, rather than be scared by it . . . and accept the lunacy of it all."[15]

Soon after the launch of Birds Aren't Real, YouTuber PewDiePie uploaded a *Meme Review* video in which he discusses the Birds Aren't Real meme. Within 24 hours, the video garnered more than 2.4 million views and 23,700 comments, many from people who believed the story. McIndoe later posted a "confession" video of an elderly actor impersonating the plan's fictional

CIA whistleblower, Eugene Price. The video garnered over 1 million views on TikTok.[16]

The movement grew. McIndoe claims there are about 50 Birds Aren't Real chapters in the United States today, whose members meet at annual gatherings that range from rallies and seminars to protests. In 2021 Birds Aren't Real adherents demonstrated outside Twitter's headquarters in San Francisco to demand that the company change its bird logo.[17] The Facebook page has 112.7K members, including the "Bird Brigade," the hardcore who believe the conspiracies.

Many of the group's adherents know the conspiracy is false, but continue to "protest" anyway, as a form of performance art. Pittsburgh-based Birds Aren't Real organizer Claire Chronis, now 23, said, "My favorite way to describe the organization is fighting lunacy with lunacy," adding that "it's a way to combat troubles in the world that you don't really have other ways of combating."[18]

Fake News

Virtual Natives know very well how to create content on social media and in 3D environments, and because they know how easy it is to create, they don't take it at face value. Unlike previous generations, such as Boomers, with their unwavering reverence for formal media channels, VNs get most of their news from social media. The difference is that VNs don't trust what they read in social media, and Boomers do.

One study of older people showed that because they had lower levels of digital literacy, they tended to believe social endorsements were based in fact simply because the information was relayed in print.[19] After looking at several factors, including political affiliation, the authors of the study concluded that age – not politics – was the key determinant in the spread of fake news.

Virtual Natives know that *they, themselves* are the source and consumers of news and, as such, they realize that "news" is subjective. In contrast, people aged 65 years and older were shown to be seven times more likely to share fake news than 18- to 29-year-olds, in a 2019 study by Princeton University.[20]

The elderly are also much more likely to fall prey to scams, online and otherwise.

The phone rang. Eighty-three-year-old Marjorie picked it up.

There was a brief pause, then a click. "Hello. Can you hear me?" asked the pleasant voice on the other end.

"Yes," Marjorie replied. The call was then disconnected. Puzzled, Marjorie hung up.

The phone rang again. When Marjorie picked it up, the same pleasant voice asked, "Who is this?" Marjorie replied, "Why, this is Marjorie Anderson. Who is this?" The phone call disconnected again.

Unwittingly, Marjorie had just handed scammers all the tools they needed to steal from her. By giving them a recording of her saying "yes" and confirming her name, she had made it possible for scammers to use her identity to make their own online and phone-based purchases.

Sadly, the "Who is this?" scam is only one of many. In the United States in 2021, there were 92,371 older victims of reported fraud, resulting in $1.7 billion in losses – and probably much more in unreported losses. The FBI has found that most common financial scams targeting older people include government impersonation scams, sweepstakes scams, and robocall scams.

Where would a VN have avoided the trap that caught Marjorie? Pretty much at every point. Scam-savvy VNs know that a pause and a click before the caller speaks usually means a robocall, and that saying "yes" and giving your name to any unknown caller means trouble.[21] Millennials and Boomers are less likely to have the general level of tech knowledge that keeps them from falling into not just this trap, but also the many more that continually proliferate as rip-off artists eternally look for ways to defraud their unaware victims.

Today, Virtual Natives are excellent at determining when they're being manipulated online, and are able to either fight inauthentic or predatory content, or play along with it if they choose – it's their choice, and under their control. Where tensions may arise is when increasing numbers of VNs enter adulthood, and begin to gain direct experience with the predatory and profit-seeking practices that have become standard in many industries, practices that earlier generations have accepted with a shrug and a sigh (and sometimes air rage). Just as we saw earlier that many VNs are starting to embrace Web3 and crypto as a way to avoid a banking system that does not truly serve them, we may well see that VNs begin not only to shine a spotlight on companies with poor practices, but also to go further, and begin to create disruptive digitally based solutions and workarounds to build the more responsive, effective services that they expect. Or at the very least, when those industries are difficult to replace digitally, they'll flock to new competitors who champion authenticity and integrity in their rejection of widely accepted, yet objectionable, practices.

If your industry has been complacent about its practices that serve its own goals at the expense of customer experience, be warned: The Virtual Natives have the tools and desire to make you obsolete.

13

Love, Sex, and Algorithms

IN THE PREVIOUS two chapters, we looked at how Virtual Natives possess the ability to outsmart systems and services that benefit the industries that created them far more than they benefit the end user. But there is a whole other class of algorithm out there that doesn't benefit the end user as well as it could – algorithms based on outdated ways of thinking that draw on old-fashioned categories of who people are, and who they could become.

Are Dating Apps Designed to Fail Users?

Some pundits posit that search engines like Google are designed to both help you find information, and thwart you from finding it. Why? The reason that you often don't *quite* find the right answer on Google is that advertisers pay to be listed on the first results page, so the best answer to your query might not appear anywhere near the top of the results. Are there other algorithms that are designed, essentially, to fail? This is a wide question that we'll explore by looking at one area in particular: online dating.

Dating apps are used by millions of lovelorn users to find their match. For Virtual Natives, love has always been found online.

"When I was in college and you went to a bar, you had to exchange phone numbers. Heaven forbid you lost that number, or you'd never see that person again," Jason, 47, reminisced. "There was absolutely no other way to find them, so losing a number was torture! Then, all these apps came along. It was like, who does this? Now, it's totally normal, everyone is on them."

Collecting crumpled scraps of paper in your coin pocket would seem crazy to Virtual Natives, whose phones are always at the ready to connect across multiple social media accounts and potentially follow your every move. But is dating really easier? What does a relationship mean today? Heck, what even is gender? And do we even need to date in human form anymore? We'll get to that. But first, we'll explore the wonderful and terrible world of dating in the virtual age.

The user experience of dating apps tends to reflect the cultural values of those who create them. Right now, most dating apps have been created by people in generations older than Virtual Natives, and it shows. For example, most of these apps want you to choose among "man looking for woman" or "woman looking for man," or "woman seeking women," or "men seeking men." It's always binary. Why is that? And why are age and height embedded into the algorithms? These are all cultural biases, baked into the UX design, which no longer serve the needs of Virtual Natives.

In online worlds, gender is optional. As we saw earlier, male professional gamers play some games as females for strategic reasons. You just can't make any assumptions – that "curvalicious" femme fatale fighter chick (think: Lara Croft) is as likely to be a G.I.R.L. ("guy in real life") as a female or AFAB ("assigned female at birth").

Younger generations don't need to make hard, existential choices about their identity. In the past, courtship and marriage was seen as a way for young people to move forward in life. Those rituals and hard deadlines are now becoming obsolete. So, too, is the need for fixed identity and self-definition, because today, the possibilities for personal self-expression are unlimited.

In the past, age has been a signifier for what a person is supposed to have achieved by a certain point in their lifetime. So when a person says "I'm 24," one thinks: You've possibly attended college, or, you're at the start of your career. But what if that same person started a successful online shop at 12 years old? If Virtual Natives don't feel the need to attend college or buy a home within a given time frame, what is "time"? In the end, age may truly become just a number, decoupled from meaning as a signifier of your likely level of life accomplishment.

VN's know they can create an internet identity that reflects what they *want* to be. They experiment with the medium, assuming any variety of people, creatures, animals, vegetables, and more, chafing at what they see as artificial boundaries and categories that made more sense in the past.

How Are Relationships Different on Virtual Media Platforms?

With their ability to see the reality of what happens in both physical and digital spaces, Virtual Natives inhabit digital worlds as naturally as they do the physical one, and easily form relationships with others in both.

Wyatt Hale, a high school junior in Bremerton, Washington, spends a lot of time on YouTube, where he connects with friends he has cultivated all around the world. Their intimacy over time has evolved to the point that he could tell you "everything about them," he said. Just "not what they look like in day-to-day life."[1]

Friendship is a natural outcome of frequent social interaction across virtual spaces. So, too, is love.

In the subreddit *R/virtual-reality*, multiple users have identified themselves as having met or begun relationships in virtual reality environments. Some people even went on to meet in the physical world and ended up getting married, while other relationships remained platonic. But all of them have formed an emotional attachment with someone that they met online. In immersive media, the setting is virtual, but the emotions and connections are very real, whether the participants ever meet in person or not. Which brings us to a confession from our own author, Catherine.

In 2020, Catherine had a weekly talk show in the Metaverse with a colleague, Michael Barngrover, called "Spatialand." The show took place in Altspace, a virtual reality app, and each week they would discuss major VR events, experiences, and media. They would show clips from final selections from the Tribeca Immersive Film Festival, Venice VR Festival, take the audience on a tour of the newly opened Burning Man VR world or on other field trips, and discuss hot topics and events. The audience was small, but international and regular. One audience member always stood out to Catherine: Mr. Carrot.[2] Mr. Carrot was charming and chirpy, and when audience discussion opened, he always had something interesting and highly relevant to contribute. It soon became clear to Catherine and Michael that this person was probably in the industry. While doing research and otherwise hanging out in VR spaces outside of the podcast, Catherine would occasionally run into him, and Mr. Carrot would take her to visit and play in new worlds. Once, she took him to her favorite Disney adventure, where they spent hours exploring classic Disney scenes, and what they jokingly called "Disney's dungeon room." After a few meetups, Catherine realized that she had developed a crush on Mr. Carrot.

She knew he worked in California, as she did, and that he probably worked in the VR field somehow. She was curious to know who he was but rationally, she wasn't tempted to actually try to find him. To non–Virtual Native Catherine, the idea of getting together in the physical world with someone whom she only knew as a gigantic vegetable online was, well, not going to happen. In her world, that's only what desperate recluses might do. It wasn't "natural." Eventually, in the spring of 2021, the world slowly opened up after lockdown. When Catherine's show concluded, she lost touch with the audience, including Mr. Carrot.

Catherine's flirtation in the virtual world was short-lived. Others, however, go on to find deep and lasting connections in digital spaces, even if they never meet in person.

"Sometimes not being able to touch or feel the person that you love," says a pink-haired avatar, "you fall in love purely with their personality." This character was captured in Joe Hunting's film, "We Met in Virtual Reality," which premiered at Sundance Film Festival in 2022. Like his previous short film on the subject, "A Wider Screen" (2019), Hunting's film explores the lives and relationships of people who had met and fallen in love on the VR Chat platform. Both films introduce a group of people whom

Mr. Hunting has met and interviewed over time, sometimes singly or in couples, and they talk about how the medium has facilitated their relationships. Many of the relationships are reciprocal; others are not.

What Hunting's films touch upon is the depth of the emotional connections formed. And while VR does enable people to portray themselves as humans who are much smarter or sexier than they are in real life, just as often they appear as mythical creatures, animals, robots, and even . . . carrots. What is striking about both films is the freedom of self-expression in the digital realm that enables characters to revel in who they are, and appreciate others on a different level. The pink-haired cartoon creature summed up the feeling of many when she expressed that "The one I love most I met in VR and I've never met them or seen their face." For many, perhaps the sensation of love created by the union of two minds is just as important, or even more so, than its physical expression. For Virtual Natives these invisible liaisons feel natural and very real.

Is Love a Game?

The documentary series *Happily Ever Avatar* follows three couples whose relationships began in the games they love, before they finally met in the physical world.³ The 12-part series, launched in 2018, explored how gaming brought couples together. It further gamified their actual relationship by seeing if couples were able to "level up" by making an effort to actually "win" this partner, or if they ended up "logging out." The show encouraged people to meet in person to see if the romance would actually work.

Samantha Hanks, the executive vice president of casting at Magical Elves, the production company behind *Happily Ever Avatar*, observed, "These people fell in love not based on what they looked like, or how much money they had, or the car they drove, or where they lived, or who their friends were. When they fell in love, it was based on the common love of these games and just talking to each other, hearing each other's voices for days, months, or years."[4]

"That's such a pure love."

The television show *Love In* took another approach to testing virtual relationships, more of an avatar-meets–Cyrano de Bergerac concept.

The showrunners chose eight lovelorn souls to use avatars of themselves to find romantic companions. But there was a catch. Their "avatars" were not digital cartoons, but other human beings, actual people. After meeting online without seeing each other in the flesh, the couples went on dates "through" actors, who voiced the answers provided to them by the contestants, who were located offscreen. The actors ranged dramatically in age and appearance, challenging the contestants to look well beyond the superficial.

Another twist on the theme of virtual dating comes with avatar dating apps. In *Avatar Life*, people can select an avatar to represent them when they go on virtual dates. *Avatar Life* is both a chat and a simulation game that encourages users to "create your own avatar and become whoever you want: a party or an upright girl: Play, chat, fall in love!"[5]

The app gives users the ability to create an ideal world in which you control everything. To make your avatar, you can select from over 800 pieces of clothing and 400 physical configurations, with the ability to modify your body shape, height, hair, lips, nose, and eyes. Once "avatar you" has been created, you can build your "dream house," choose a "job," and earn in-game currency. The app encourages you to "Throw a wild party in your apartment, treat your friends with drinks, and dance through the night. Don't hold back and throw a cake in someone's face!"

Given the fantasies suggested by the *Avatar Life* app, it's probably a good thing that no one knows who you are.

Over in the digital world *Second Life*, the Lonely Hearts Dating Agency, established in 2010, is one of the virtual realm's oldest and most successful dating agencies. They host numerous events for avatars to meet and seek potential partners. One such venue is Romance in the Clouds, where you can "enjoy slow ballads ripe with passion as you dance in your lover's arms. Romance in the Clouds is always open for your pleasure."[6]

Despite being a relatively new concept, the longevity of the Lonely Hearts Dating Agency suggests that avatar dating fills a genuine need in the world of emotional connection. Meeting someone else while both you and they are avatars frees you both up from whatever labels you may bear in the physical world, giving you the space either to play at being someone or something other than you are, or to be the most authentic you possible, unencumbered by your physical form.

Fortnite and Flirt

If "Netflix and Chill" was the Millennial mantra, what is it for Virtual Natives? As it turns out, the answer is gaming. Seventy percent of Gen Z gamers spend an average of seven hours per week in online games like *Fortnite*. But they're not always there to play the game, according to research. They're there to socialize, which they do just as much as they game.[7]

So instead of "Netflix and Chill," the mantra for Virtual Natives may be "*Fortnite* and Flirt." Because, let's be honest; while Virtual Natives may feel comfortable with their sexuality, they aren't having much sex.

Virtual Natives may be having less sex because Covid prevented people from meeting in person for a couple of years, which may have forced some activity online. In a 2022 study by Lovehoney, 25% of 18- to 24-year-olds said they've never had partnered sex. Forty-six percent of the same age group, or nearly twice as many, said that they have engaged in some type of virtual or online sexual activity – but they don't count that as "having sex."[8]

Overall, younger generations show more hesitancy than older generations when it comes to physical hookups. In the same survey, the Virtual Native age group was by far the least sexually curious or adventurous, partly because they felt they couldn't trust that what a partner told them about their health was true. It seems that VNs bring the same skeptical approach they use in their evaluation of social media into the bedroom when it comes to potential partners.

While the impression may be that Virtual Natives are surrounded by sex online and generate and share titillating content themselves, this does not necessarily translate to their values or practices when it comes to their actual bodies. They are physically partnering less, and spending more time, online and off, on activities that are both more social and less risky, with a steadfast commitment to their overall mental health and well-being.

Love Is Love

From afar, it can seem as though the world of Virtual Natives is brimming with sex and infinite possibility. In practice, though, it seems that VNs have reacted to the duplicity and uncertainty of the modern world by moving away from physical sex and toward a more "pure" meeting of the minds and emotions in digital worlds. Being represented by an artificial avatar in situations in which true relationships and connections are formed exemplifies many of the themes we earlier identified as vital for understanding Virtual Natives: the interchangeability of the digital and the physical, their ability to willingly embrace new identities in themselves, and their ability to willingly embrace new identities in others.

Their attitude toward physical sex may be more conservative than that of their elders, but their attitudes toward categories such as gender and what is represented by calendar age are certainly more fluid. This brings us back to the structures behind dating algorithms and their baked-in assumptions about cultural "norms," which are expressed in traditional categories such as gender preferences (male seeks male, male seeks female), height, and age definitions. For Virtual Natives, much of this won't make sense.

And it's not just dating apps – we focused on them to make a wider point. Any kind of service or situation that attempts to classify people based on rigidly drawn categories of any kind is going to find it increasingly difficult to wedge its customers, employees, or even citizens into tidy little boxes. And, as we'll see in the next chapter, that becomes an even thornier issue when we look at race.

14

Web3 and the Culture

IN THE PREVIOUS chapter, we pointed out that traditional approaches to personal characteristics such as gender and age may prove to be a poor match with how VNs see themselves, given their openness to identity fluidity and willingness to follow alternative life paths made available through digital tools.

The question of race for VNs is both the same, and different, as the question of gender. It is the same as gender in that it is no longer as clear-cut as it used to be. The 2020 US Census revealed that the number of people who identified themselves as neither one race nor another, but rather "multiracial," rose from 9 million in 2010 to 33.8 million in 2020, a 276% increase.[1] But no matter how blurry the lines between male and female become, there are still only two main starting points when you're talking about gender: male and female. Overall, and acknowledging that great advances need to be made in representation for nontraditional variations on these two categories, both male and female people are pretty well represented in popular culture, training data for algorithms, and so on.

It's a different story when we talk about race. There, we have an infinitely long list of possible categories, including every race and ethnicity on the planet, and every possible combination of two or more of them. Out of all this glorious mélange, people who identify as white are still in the majority within the United States, as of the 2020 census[2] – and it shows. White is the default that tramples the visibility of other races, from who gets to headline in commercial media, to whose culture is deemed the norm, to whose photographs are used to teach computers what "human beings" are supposed to look like. People who identify as white are projected to no longer be the majority in the United States by 2045 or so, a short couple of decades away.[3] VNs are aware of this, and are looking for representation and understanding that better matches their mixed-race, multicultural heritages, that authentically sees who they are and celebrates them for it. It's the future of the country, after all.

Whole Lotta Love

Culture is the glue that holds communities together, whether in the physical world or online. In American culture, many music and dance genres that later became big mainstream (= white) hits were first created and celebrated by minority groups, Black people in particular. Just as the blues gave birth to rock and roll, and later hip hop, music has been a cultural vehicle for storytellers who captured the public imagination at large and came to dominate and influence the general culture. In a similar vein, AAVE, or African American Vernacular English, has similarly become part of the fabric of youth discourse, with a range of terms such as "yo, bro, sup?," "what's good?," "finna," "hit me up," and "periodt" – just a few of many, many examples. Online, Black Twitter and Black TikTok creators have driven both the discourse and the creative content that helped to fuel those platforms, by driving both eyes and advertising to them.

> Every day we get on their platforms (TikTok, Twitter, etc.), and we create trends, music, art, and language that they turn into billions.
>
> —NoName (Fatimah Nyeema Warner, artist)[4]

If imitation is the sincerest form of flattery, then Willie Dixon should have felt very flattered. Born William James Dixon on July 1, 1915, in Vicksburg, Mississippi, a song that he wrote has been ranked Number 75 on *Rolling Stone* magazine's list of the 500 Greatest Songs of All Time. Only it wasn't attributed to him. And it was no longer considered a blues song, but a rock and roll classic. What happened?

> The Blues were born in a thousand sharecropper fields, humble shacks and rural juke joints.
>
> —VTUSA[5]

Known for its lyrical form, call-and-response pattern, and emotional style, the blues were born as a folk music, popular among former slaves living in the Mississippi Delta, the flat plain between the Yazoo and Mississippi rivers.[6] By the 1960s, this simple, expressive form of music rose to become one of the most important influences on the development of American popular, or "pop" music, including jazz, rhythm and blues, rock, and country music.

"I think America concedes that [true American music] has sprung from the Negro," said W. C. Handy,[7] who was widely renowned as the "Father of the Blues." He was the first African American music publisher whose songs, like "St. Louis Blues," went on to become classics that informed new music styles for many generations.

> When we take these things that are our own, and develop them until they are finer things, that's pure culture.
>
> —*W. C. Handy*

In August 2014, Willie Dixon's guitar riff in the 1962 release of his song "You Need Love" was voted The Greatest Guitar Riff of All Time by the BBC. Except he didn't get any credit for it, because the song that actually won the award was Led Zeppelin's "Whole Lotta Love," as played by guitarist Jimmy Page.

Much like Virtual Natives who publish their work online only to watch it take off when other, more famous, creators adopt it, Willie Dixon's original song, featuring Muddy Waters, had been a minor hit in 1962, but it hadn't crossed over to mainstream audiences and was only known to the relatively small audience of blues fans.

A comparison of the music and especially the lyrics of Dixon's "You Need Love" and Led Zeppelin's "Whole Lotta Love" makes it pretty clear that Jimmy Page was aware of the prior work when he "wrote" the latter in 1969.[8] It took until

1985, when his daughter first played the song "Whole Lotta Love" for him, for Willie Dixon to become aware of what Led Zeppelin had done. He successfully sued the group for plagiarizing his composition, and eventually won the right to be credited as one of the songwriters.[9]

Ultimately, both Jimmy Page and Robert Plant openly admitted to stealing music on multiple occasions. Robert Plant blithely considered this part of being a musician, saying, "You only get caught when you're successful. That's the game."[10]

"Knocked out" when he heard the news that "his composition" had won The Greatest Guitar Riff of All Time award, voted by BBC Radio 2 listeners, Led Zeppelin guitarist Jimmy Page commented, "I wanted a riff that really moved, that people would really get, and would bring a smile to their faces."[11] Except . . . it hadn't been his creation at all, and he still got public credit for it nearly 30 years after Willie Dixon had successfully shown that he was the actual composer. Cultural theft casts a long shadow, and one that is difficult to erase.

Do the Renegade

In September 2019, a 14-year-old Atlanta girl named Jalaiah Harmon posted an Instagram video of herself and her friend Kaliyah performing a dance she had invented, called the Renegade.[12] A popular TikTok dancer, @global.jones, brought it to TikTok the following month.[13] Within six months, everyone from Lizzo to the Kardashians were doing Jalaiah's moves. And then came the big break: Jimmy Fallon had seen "Renegade" and wanted to showcase the new moves on his show. He called newly minted TikTok superstar Charli D'Amelio to perform.

At the time she uploaded her own video doing the Renegade, Charli, like Jalaiah, was a complete unknown. It was Jalaiah's dance that catapulted Charli to fame, earning her over 118 million followers, and garnering her a spot on one of America's most watched television shows. News of her appearance spread across multiple magazines, from *Teen Vogue* to *People*,

Vox, and *Vice*. From there, she would go on to receive lucrative advertising and sponsorship deals, and widespread fame that continues today. Great for Charli, but – it wasn't her dance. It had been created by Jalaiah. Jalaiah never received any credit or thanks either from Charli or from Addison Rae, another massively popular influencer who was invited to the same Jimmy Fallon episode to "teach" fans how to dance to songs. Addison Rae's precipitous rise to stardom – in 2020 she was the highest earner on TikTok[14] – was also largely due to the Renegade. Both Jalaiah and @global.jones are Black, while Charli and Addison are white.

Sunny Hostin, host of daytime talk show *The View*, has been very direct in her criticism of this recurring phenomenon. In 2021 she said, "We're seeing it bigtime now on TikTok with black creators. They create these incredible dances. They go viral, like the 'Renegade' dance, 'Savage' dance, the 'Gitup' challenge, and then you see white teenage women misappropriate it, and they make millions of dollars off of it."[15]

Jalaiah, for her part, was initially pleased by the virality. "I was happy when I saw my dance all over," she said in an interview with the *New York Times*. "But I wanted credit for it."[16] The *Times* called her "the original Renegade," presenting her as "part of the young, cutting-edge dance community online that more mainstream influencers co-opt."

But today, Jalaiah isn't an influencer. She doesn't have any brand deals, hasn't received any media opportunities, and is largely ignored by the professional dance and choreography community. It hurts.

"I think I could have gotten money for it, promos for it, I could have gotten famous off it, get noticed," she said. "I don't think any of that stuff has happened for me because no one knows I made the dance."

Sometimes the creators strike back. In 2021, Megan Thee Stallion released a new song, "Thot Shit." On TikTok, no one danced to it. Not because it wasn't a danceable song – it was – but because Black creators on the platform had become tired of coming up with new dances, but having white creators get the credit, the followers, and the money, for their renditions of the same dances. So they went on strike, and no one created a new dance to go with the new song. Without a dance to go with it, the song did not go viral, unlike all of her previous hits.[17] Using the hashtag #BlackTikTokStrike, Black creators thus voiced their objections to what they saw as preferential treatment of white content creators, whose work was more often picked up and promoted by mainstream media, earning those creators millions in the process.

White TikTok creator Rachel McKenzie sees the injustice, and publicly supported the strike. "Anyone that uses TikTok will tell you that young black creators choreograph the vast majority, if not all, of the dances that go viral," she affirms. "If you look at modern pop culture and its entirety, it's just another example of how black culture [is popular], and white people hijack it."[18]

When the Default Is the Same Type

While this unfortunate pattern of one group creating and another group copying and getting credit continues on widely available and accessible platforms like TikTok, one bright spot is that at least minority groups have enough access to smartphone-based media to be able to create content there.

The cultural lift that the Black community gave to Twitter, another widely accessible smartphone app, is well documented. From memes to conversations, "Black Twitter" is one of the most powerful cultural forces that propelled the platform into becoming a vibrant, compelling community, with content to match.[19]

But there are other domains that have remained far less accessible to Black participants. A different area that lacks adequate minority representation is within the actual employee ranks of the companies that build many of the digital tools used by Virtual Natives. While Black people make up 13% of the US population, in 2021 they were estimated to constitute only 4.4% of employees in Silicon Valley tech companies.[20] Part of this may be due to the digital and educational divide within the United States. Significantly lower internet availability rates and a lack of access to new technology and skill sets is the unfortunate reality for many VNs from minority backgrounds, preventing them from playing roles as either participants or creators in immersive worlds. And it shows.

Modern computing's lack of multicultural understanding is something that makes itself clear in many small ways on a daily basis. For example, Maya, a Black American, was experimenting with the Generative AI platform Midjourney to create a visual avatar for herself. As is common when working with Generative AI platforms, she started by entering a prompt that consisted of a written language description of what she wanted. Her first prompt was along the lines of: "Futuristic female, staring at Metropolis at dawn. Cinematic, highly realistic." Midjourney created a batch of responses that looked cool, but – the images were all of a white futuristic female. When she further specified, "Black," the images instantly became dark-complexioned, which Maya is not. So she added "Rihanna" as part of the criteria. White again! She racked her brain for a popular celebrity of the shade that she was after that the AI might recognize. No luck.

At this point, she realized that the paucity of highly success-ful, popular Black actresses and role models probably explained why the algorithm struggled to generate various types of people. There just weren't enough of them, or not enough that were suf-ficiently famous to influence the algorithm's data set. No wonder it was defaulting to Caucasians, even when she included famous non-Caucasian women in the prompt. So frustrating! Midjour-ney, and other Artificial Intelligence platforms trained on pri-marily Western, Caucasian data sets, hasn't gotten the news yet about the United States' rising multiculturalism.

What's deeply ironic about Maya's experience is that the digi-tal realm is meant to be where you can be whatever you want, as we've mentioned earlier. Unfortunately, it turns out that if you want to be a platypus or giant omelet, that's fine, but if you want to accurately represent your non-white identity, the algorithms are just not going to be able to help you. That's a problem. As the old adage goes, "It takes all kinds to make a world." We would be bored if there were no other countries or cultures to be discovered, if we all wore the same clothes, hairstyles, ate the same food, and listened to the same music. Should online worlds be any different?

LaJune McMillian ran into the same issue in 2016 when looking for reference images for her 3D dance experiences developed in Unity. A young Black multidisciplinary artist, teacher, and creative technologist, McMillian discovered that it was basically impossible to find models of bodies like hers in motion that she could bring into 3D platforms. So she decided to build her own database. Over the following years, she set out to capture data from Black performers and create Black character base models that were underrepresented in available online databases. Her final performance pieces capture a dancer's movement and project it simultaneously on-screen with AI-generated imagery, while a voice recording recounts the dancer's thoughts, life experiences, and emotions as they move. It's a powerful combination, and only possible when the AI models that she uses accurately reflect the people and culture that she honors.

Economic Access Is Key

Virtual Natives, 48% of whom in the United States consider themselves to be more than one race or ethnicity, will, like Maya, want at least one of their avatars to look like them. But even more important than seeing themselves online, they will want the keys to the economic engine. A more diverse creator pool will not only enrichen the ecosystem, but will also open up important economic opportunities for a broader population of

Virtual Natives. And that should make virtual spaces far more interesting for everyone.

We have addressed gender and sexual orientation multiple times in this book, but accounting for racial, ethnic, and socio-economic diversity is of equal import. Indeed, McKinsey states that inclusion and diversity are "critical for resilience, and reimagination." In other words, innovation: "Diverse companies will out-earn their industry peers," they conclude in their report "Why Diversity Still Matters."[21] Their findings demonstrate that diversity drives both organizational health and better business performance. In fact, they identified a substantial performance increase – of nearly 50%! – by companies that are highly diverse, compared to the least diverse companies. Yet their findings also showed that some industries, including many of technology's biggest leaders, have made little or no progress toward realizing true diversity and inclusion – and some have even gone backwards since tracking began.[22]

Virtual Natives are natural creators, and as a generation they are not only highly diverse but also highly inclusive. They will expect the companies they work with to reflect those values. Says Yoni Wicker, 14, of TheWickerTwinz, "We have 1.7 million followers and we always give credit whether the person has zero followers or not. We know how important it is."[23] The relatively anonymous creators who inspire influencers want the world to acknowledge their work, and by extension their worth. "That person who made that dance, they might be a fan of ours," Yoni says. "Us tagging them makes their day." And who knows? It might even be the beginning of a whole new career.

Culture is a thing to be shared and celebrated. The increasingly multicultural cohort of Virtual Natives has no problem creating and sharing digital content; it's more natural to them than writing by hand. The current problem for minority creators lies in the flawed honor system for giving credit where it's due, as well as in the biases of the larger culture, especially traditional media, that repeatedly select Caucasian ambassadors over other ethnicities when they want to hear from Virtual Natives. Another serious issue is the inherent bias of the digital tools used by Virtual Natives, created by groups of people with a very different demographic mix than that possessed by Virtual Natives.

Just as Virtual Natives are growing increasingly intolerant of corporations that focus on profit over providing services – and are thus focusing on alternatives based in Web3 and crypto to take back their financial power – those who continue to be underrepresented and even misrepresented are likely to embrace disruptors who enter the space with products designed to fill these representation gaps. Given the VN preference for diversity, authenticity, and acceptance, they are likely to be grateful toward institutions who genuinely see them as they are, and who supply tools, services, and recognition that match that reality. This gratitude will be matched by the VNs' love, attention – and spending power.

15

Life Is Games

LET'S TURN NOW from the topic of underrepresentation and the powerlessness that it represents, to an area in which Virtual Natives have much more agency: gaming.

When we looked earlier at the relationship of Virtual Natives and their sense of personal agency, their feeling that they should be able to use their digital tools to take action and solve problems, we mentioned in passing that gaming had something to do with this mindset, in that it creates a sense that if VNs make the right moves, they should be able to level up, in life as well as in their games. We talked about this topic earlier in relation to the workplace, but let's look at it again with a wider lens. After all, 90% of Virtual Natives consider themselves to be gamers or game enthusiasts, so the gaming experience is a central part of their lives.[1] What psychological expectations does a gaming-heavy childhood create in VNs, and how does this experience set them apart from earlier generations?

"Red Light, Green Light, One, Two, Three!"

The mechanical doll turned around.

She looked like a schoolgirl of about eight years old with neat, short black hair pulled back into ponytails on either side of her head. Her orange tunic playdress was layered over a yellow blouse with a princess collar, and she wore long white knee socks. It was the kind of outfit you found on dolls made before the 1960s, but her cuteness was diminished by her gigantic proportions. At some 20 feet tall, she towered over the uniformed group assembled in the gymnasium before her.

"Stop!" she commanded.

Hands out, torsos swaying, knees and ankles wobbling, most of the players came to a halt. The others, caught midair by surprise, thumped belatedly to the ground. Shots fired. Those who did not completely stop in time were now on the floor, bleeding, dying. The doll had shot them from automatic rifles that had, emerged from holes in her eyes.

Corpses now lay all across the floor. The remaining uniformed men and women, now desperate, could not escape. The doors were sealed. The *Squid Game* had begun.

Fictional survivalist dramas have been at the forefront of culture since the 2010s, ranging from the dystopian futuristic survival adventure *Ready Player One*, in which a young man seeks to win a game developed by a mythic gamer by competing against corporate interests that effectively enslave half of humanity, to the *Hunger Games*, an annual event in which a boy and girl between the ages of 12 and 18 are randomly selected to compete to the death with rivals from other districts for the amusement of the economic elite. These dystopian battles for survival have been joined more recently by *Squid Game*, in which desperate men and women risk their lives in a soul-crushing game that will deliver one of the 456 people from their worldly troubles, but all the rest will die.

Each of these films poses important questions to today's youth: Is life a game, or a kind of reality show? If so, what would you do to survive?

What *wouldn't* you do?

This is a compelling, and universal, question. So universal, in fact, that it propelled *Squid Game* into Netflix's top viewership ranks, a first for a South Korean–language television series. The simplicity and relatability of *Squid Game*'s deadly challenges have also made it a popular theme for VN creators across gaming platforms such as Roblox and Minecraft. There fans can play the games for themselves, or, after they've been killed off a couple of times, watch professional VN gamers stream their own efforts to beat the system on Twitch and YouTube. Popular examples of *Squid Game* survival attempts have racked up hundreds of thousands and even millions of views.[2]

Squid Game creator Hwang Dong-hyuk explained, "I wanted to write a story that was an allegory or fable about modern capitalist society, something that depicts an extreme competition, somewhat like the extreme competition of life. But I wanted it to use the kind of characters we've all met in real life."[3] He wrote the story when he himself was struggling economically, which made it all the more personal and potent.

Like other survival dramas, including *Hunger Games* and *Ready Player One*, *Squid Game* is a cautionary tale that examines the extremes of social power, injustice, and privilege, and the existential threat that polarization poses. These are big, serious issues, which earlier generations tended to explore in literature or film, or both, for example, in *Fahrenheit 451* and *1984*. To be fair, both *Hunger Games* and *Ready Player One* also started out as hugely popular books. But Virtual Natives are taking these darkly themed works about the imbalance of power, and turning them into candy-colored games. Why are they building these games themselves in Roblox? Because a commercially produced studio version isn't available and, as fans, they want to live the experience firsthand. So they have built the game that they want to play. And once they're done playing their homegrown versions of *Squid Game*, there are plenty of other survival games out there to choose from.

One example is *Rust*, whose objective is simple: "Do whatever it takes to last another night." The goal is survival against a brutal environment, other inhabitants, and the climate itself. *Ouch*. As *PC Gamer* enthused, "*Rust* is one of the cruelest games on Steam, and that's what makes it so compelling." It also contains "violence and caveman themed nudity."[4] Say what, again? Interesting.

Fundamentally, Virtual Natives seek challenges in media that they are lucky enough not to be faced with in real life. The likelihood of being bitten by poisonous snakes, being stranded in a dark, frozen icescape, having to traverse sweltering canyons while being pursued by large raptors, or being forced to play a lethal version of marbles is unlikely, to say the least. If we're honest, many of us would struggle to start a fire without matches.

There are numerous TikToks dedicated to the subject of what it would take to survive the *Hunger Games*, many of which point out the poor odds of winning. One would need to be very athletic, have combat experience, forge successful alliances, build shelter, and identify and avoid deadly flora and fauna, not to mention one's opponents. But wouldn't it be amazing if you did have all of those skills? If you could meet these challenges, and overcome them?

It's in games like Minecraft, *Rust*, Roblox, *Among Us*, and *The Maze* that Virtual Natives are asking those questions of themselves, and finding out the answer. These aren't just strategic shoot-'em-ups, they're challenges of ingenuity and will, cunning and strength, tests of both mind and body. In fact, VNs are testing themselves in survival games in the same way that their ancestors might have done during an initiation ceremony to earn one's ascendance to adulthood.

In many ways, modern-day survival games resemble ancient coming-of-age rituals that adolescents performed around the age of puberty in order to be considered an adult by the tribe. In one of the most extreme examples, Australian Aboriginal elders would teach children essential survival skills until about age 11–16, when the child would be sent to the wilderness armed with nothing more than their wits, a loincloth, and some body paint. The child would then do a "walkabout," or wander alone for six months, foraging for his food and water, and making his own shelter to prove his self-reliance in all situations.

Luckily for today's Virtual Natives, the real fun of survivalism is that it's virtual. They play these games to imagine how they might use their skills and intelligence to survive, with the definite bonus of knowing they won't get hurt, even though some of the games are indeed quite violent. This brings to mind another traditional media format that has the same mix of violence, children, but nobody actually coming to harm: fairytales.

For instance, *Hansel and Gretel* is a terrifying story of child neglect, abuse, and cannibalism. Similarly, the fable *Der Struwwelpeter* ("shock-headed Peter"), also recorded by the aptly named Brothers Grimm, tells the tale of a boy who refuses to cut his fingernails, only to be visited one night by Struwwelpeter, who solves the problem by cutting off *all* the boy's fingers! Or *The Juniper Tree*, in which yet another evil stepmother kills her stepson and serves him to his father as a stew. The list of frightening children's tales goes on and on.

Why are fairytales brimming with dark forests, evil stepmothers, abandonment, cruelty, and arbitrary violence? In his classic work *The Uses of Enchantment: The Meaning and Importance of Fairy Tales*, psychologist Bruno Bettelheim says the answer lies in a "child's need for magic."[5] These stories engage our subconscious and speak to our darkest, most potent fears and anxieties – and our need to confront them to survive.

Bettelheim says that myths and fairytales answer fundamental questions such as "What is the world really like?" and "How am I to live my life in it?" The child learns through the stories they are given, but through the prism of magic. Before they hit puberty, children struggle to distinguish between live and inanimate objects, often giving both agency, imagining that birds and trees, for example, can talk.

As represented by Damien Hirst's artwork, a pickled shark in a glass tank entitled *The Impossibility of Death in the Mind of Someone Young*, Bettelheim says that children have difficulty distinguishing between living and dead things. Having been very much alive with not much memory of their physical existence, children can easily imagine that dead things might easily come to life. Explaining that the earth is held aloft by gravity and speed is, to a child, just as valid as saying it's being held up by a giant, or balances on the back of a tortoise. Similarly, repulsive wart-faced crones can become kind accomplices, just as beautiful queens can transform into wicked, poison-bearing stepmothers.

Fairytales and myths have held an important and enduring role for children around the world and across the centuries because they provide children with the tools to imagine an empowered way of being in the world, despite many hardships. These tales help children imagine that with wile, cunning, courage, and a dash of luck from helpful allies along the way, they too may triumph over any adversity they face – and make even the most unbearable life worth living.

Fairytales are still with us, because they still serve their original purpose. But take a look at this description of the heroic fairytale protagonist from Bettelheim:

> The fairy tale hero has a body which can perform miraculous deeds. By identifying with him, any child can compensate in fantasy and through identification all the inadequacies, real or imagined, of his own body.[6]

Doesn't this sound an awful lot like gaming? With their hands on the controls of their gaming consoles, VNs literally have control, and, like Grimm heroes of old, they too can climb through the clouds, outwit giants, battle winged beasts, or change their appearance and become the most powerful or beautiful or best-loved person in all the land. By satisfying those needs in fantasy, or in the digital reality of a game, the child can be more at peace with their world.

"Matsya Nyaya" is a Hindu phrase that means "The Law of Fish." As a rule of nature, the small fish become prey to big fish; the strong devour the weak. To survive, the little fish must become more clever and agile if they want to survive. For children, myths have traditionally helped them achieve that goal. Today, gaming serves the same purpose. You don't have to be strong if you can be smart.

In their desire to take on and slay the figurative dark beasts of the forest, VNs are no different from the hundreds of generations that have gone before them. What is different is that VNs are able to slay those beasts not just in the figurative sense, but in the literal sense as well, through challenging games of all kinds. This is empowering, and reinforces the sense of agency and ability to take action that VNs have developed through their creative use of digital processes in general, as discussed earlier. Because they see reality as a continuum that links the digital and the physical, their accomplishments in the digital world have increased significance in the physical world. It's not that they in any way confuse what they accomplish in the digital world with physical reality; it's more that the confidence they gain through digital successes carries over and gives them that extra nudge of confidence and self-assurance even when they're away from the keyboard. They've got this.

Where this takes on additional relevance for those who work with VNs is in the understanding that not only are VNs accustomed to taking action, but they are the literal heroes of their own lives. Instead of reading about other heroes saving the princess or slaying the dragon, they themselves have done both, maybe in *Legend of Zelda.* And if they failed the first time they went up against the dragon, they knew that there was probably another, better sword they could earn that would give them a better chance of success next time. In other words, they've been trained by playing games that (1) there is a path to success, (2) if at first you don't succeed, try, try, again, and (3) if you're really stuck, there's probably an ally nearby who can help you if you just look for them. Above all, they expect to have agency and to be able to actively solve problems. Every game is winnable – eventually – if you can just find the right combination of skills and perseverance.

Note that "agency" is not the same as "power." It's not that Virtual Natives expect to get everything that they want, right this very minute – they're not spoiled, or mini-despots. On the contrary, gaming has taught them to work for what they want, and to build up their resources until it is the right time to strike. All they expect is the ability to make their own choices and act accordingly.

Given all this, what is a sure way to torture a VN, and frustrate them beyond belief? Take away their phones, yes. But also being in a situation in which they are powerless is very hard for Virtual Natives. Being in a situation in which there is no interaction, no ability to shape the course of events, only observation. Or a situation in which they are unable to score points or earn credits. Or in a world where advancement is impossible, no matter what action they take.

The popular phrase "I feel like I should be able to" is central to VN culture. It implies a presumed expectation of agency, and their natural rights to assert their positions, opinions, and creative expression within the world.

For VNs, a situation that is a game that can't be won is a game that isn't worth playing.

16

The Charismatics

WE SAW EARLIER that Virtual Natives have shown themselves to be more successfully resistant to misinformation on social media feeds than their elders, through a high awareness of how digital everything works, plus finely tuned antennae for detecting the inauthentic. We have also seen how they like to take the reins where possible, and chafe against arbitrary limitations on their actions. How are they likely to react when they begin to encounter the inefficiencies and helplessness associated with modern bureaucracy, especially systems still based on ancient and outmoded technologies? How will they expect the tools that they're so familiar with to be used to improve services and prevent grift?

It's early days, but some VNs are already exploring how AI, blockchain, and Web3 can bring needed efficiency to official channels, and unravel deceit before it happens. Innovation in this space is being driven by Virtual Natives, and there will be pitfalls for companies that fail to keep up.

If AI Invented Politicians

Prompt: "Describe the background of the perfect American candidate for office."

George Santos was just 34 years old when he was elected to the US Congress in 2022 to represent New York's 3rd congressional district. Dark-haired, with a boyish face and thin brown academic glasses, he cut a fresh silhouette against his august opponent, Robert Zimmerman, who, at a silvery 67 years old looked like the epitome of the Washington establishment. Santos defeated him by an 8% margin. Indeed, his was very much the story of the American dream. George Santos ran as an openly gay candidate as well as "a seasoned Wall Street financier and investor," according to his online bio.[1]

A first-generation American, George Santos was the proud grandson of Jewish immigrants who had fled persecution in

193

Ukraine, then, when the Holocaust loomed, to the United States. He grew up in a "basement apartment" in Queens, where his parents raised their family "on the foundations of life, liberty, and the pursuit of happiness." George Santos attended the elite private prep school Horace Mann before matriculating at Baruch College, where he earned a bachelor's degree in economics and finance in 2010. He went on to work at Citigroup as an associate asset manager and later, the prestigious investment bank Goldman Sachs donated money to an animal rescue group he founded. He did well and he and his family began to invest in real estate.

Only none of this existed. None of these things happened.

According to an exposé by the *New York Times*, George Santos never attended Baruch, never worked at Goldman, never worked at Citigroup.[2] He had worked briefly at a company called Harbor City, which was known to lure prospective investors with YouTube videos and promises of double-digit returns. The SEC ultimately filed a lawsuit against the company and its founder, who were accused of running a $17 million Ponzi scheme.[3]

Santos had claimed in the course of his campaign that during the pandemic he and his family had not received nearly one year's worth of rent from the tenants of 13 properties that they owned. "Will we landlords ever be able to take back possession of our property?" he asked on Twitter. "My family and I nearing a 1 year anniversary of not receiving rent on 13 properties!!! The state is collecting their tax, yet we get 0 help from the government. We worked hard to acquire these assets."[4]

Not only did Santos *not* own any properties that he had "worked hard to acquire," but he had also faced multiple evictions as a renter, and had been fined over $12,000 in a civil judgment. At the time of his election, George Santos was living in his sister's basement.

Even his animal rescue group, Friends of Pets United, didn't exist.

It's almost as if someone had prompted ChatGPT to "create a backstory for the ideal American candidate," and Santos had cut and pasted the result straight into his campaign brochures.

Naturally, after the exposé, the press wondered how no one had known about these many falsehoods. Pundits wrung their hands, asking, "How did this happen?" and, "Didn't Zimmerman do his oppo research?" How did they not know that this man was not only lying about his academic and job history, but that he also had multiple aliases? As "Kitara Ravache," Santos performed as a drag queen in Brazil; as "Anthony Devolder," he claimed he was famous as a gay beauty pageant winner, and had acted in a movie with Hollywood star Uma Thurman. How did the campaign opposition not know *any* of this?

"This whole story is completely true. Except for all of the parts that are totally made up." That's the tagline for the Netflix show *Inventing Anna*, about another famous grifter of our times.[5] Like Santos, Anna Sorokin posed as someone else, with a new name and a fictional background that included faraway Disney-like castles in her fictional motherland, Germany, and a fantasy fortune, she claimed, of more than $60 million. Anna managed to become a Citibank private client by handing over fake checks, while withdrawing $89,000 in real money from them in the process.

The truth of the matter is that the public is supposed to do its own research. But in an increasingly busy, noisy world, it's not always that easy.

Anna Sorokin somehow managed to charm and bamboozle Citibank out of actual cash, indicating a worrying porosity of the banking system, whereas, if you or I want to get something like a copy of your own birth certificate, you'll find that the bureaucracy involved will nearly kill you.

If you were born in Manhattan, as both the authors of this book were, and you need a copy of your birth certificate, you can order one in a large grey marble building at the foot of Manhattan Island. It requires literal hours of Kafkaesque IDs, forms, and lingering in multiple municipal corridors and, of course, a fee. Ultimately, you'll receive a flimsy piece of blue paper with flowered etchings around the corners, announcing the date of your birth as certified by the state.

In the age of computers, how is it that we still require all this cross-documentation plus a half a day of waiting, in person, to obtain proof of our very existence?

What if you didn't have to order school transcripts, birth certificates, or confirmation letters from your employer? What if there were a technology that would unassailably hold all these things for you, available for consultation whenever you decide to release the information?

What if we had it already?

Blockchain technology is most often associated with cryptocurrency, but the two are actually separate technologies. Cryptocurrencies are just one of many information types that are supported by the blockchain. Blockchain is a digital ledger, like a virtual filing cabinet, where such information as your birth certificate, social security card, passport, drivers license, personal IDs, diplomas, employment history, medical history, and anything else can be stored. Each one of those items is stored as a "block" which, like a file, can be pulled up and shared with people such as a prospective employer, or an electoral board, at any time you wish.

A few bold souls have already started using blockchain technology to store the personal information of their families and themselves. In one famous early example, entrepreneur Santiago Siri stored the record of his daughter's 2015 birth on the Bitcoin blockchain. He was interested in finding a way to create an official, undeniable record of the fact of his daughter's birth, yet one that existed independently of official stamps from outside entities. He described the blockchain birth certificate that he had created as "a simple piece of data that can always be proven anywhere in the world with an Internet connection. . . That's incredibly powerful if you think about it. While states rely on closed bureaucracies to support their institutional belief system, the world has the Internet and the blockchain."[6] Given that the systems are not yet in place anywhere in the world to recognize the validity of blockchain-based birth certificates, Siri's action is more of a statement than an actual practical move. However, the call has been sounded: Virtual Natives see the inefficiencies of existing government-based identity validation systems and are looking for a better way to do things.

For now, most of us in the United States still sit helpless for hours in large grey buildings, waiting for our number to be called. Future generations, however, should have all their information, from birth certificates to transcripts, from employment records to fingerprints, on file somewhere. The difference will be that they may own this information themselves, rather than having a government hold it in trust for them.

Today, many people in the United States balk at the idea of centralizing data because "freedom." But what if owning your own data is the ultimate freedom? Freedom to control who sees it or can sell it is freedom indeed. There are models today that prove that such a system can work, and very efficiently. Take Estonia, for example.

The Estonia Model

A small Baltic nation with a population of just under one million people, Estonia rose to international fame in the 1990s by taking on one very focused task: becoming a digital-first society. The country, which gained its independence following the collapse of the Soviet Union in 1991, was then relatively poor. Wood, paper, and textiles were its major exports. But the government had big dreams of using technology to join the international stage.

The challenge was how to modernize an entire government apparatus that had languished for years under Soviet inattention. Rather than try to emulate centuries-old public service systems from Western Europe, Estonia decided to use the opportunity before them to leapfrog into the future.

The government appointed a young generation of entrepreneurs to create an entirely new public system, driven by digitalization. In 1997, a project called *Tiigrihüpe* (the Tiger Leap) was launched to increase national computer literacy levels. It provided computers and internet access for schools, along with digital training for teachers. The country also pledged to put computers in every classroom and by 2000, every school in the country was online. At the same time, the government undertook a massive effort to make all public services available on the internet. Healthcare, education, judicial records, taxation – they were all included. The third leg of the triangle was consciously creating an infrastructure in which internet entrepreneurs, the early dot-coms, could thrive. Pretty bold thinking for 1997.

Today, 99% of Estonia's public services are available on the web 24 hours a day, nearly 50% of citizens vote via the internet, and taxes are completed online in under five minutes.[7] And that's just the beginning.

In Estonia, all citizens are equipped with a digital ID which has multiple purposes. As a "digital ledger," it contains a personal ID, the Estonian passport, the national health insurance card, and credentials that enable the holder to do internet banking and i-Voting, sign records with their digital signature, refill e-prescriptions, check their medical records, submit tax claims, and more.

What does this enable in practice? Imagine that an Estonian man, Aksel, is biking down the street in the capital city of Tallinn, when suddenly he is struck by an automobile. Luckily, a witness immediately calls for help. An ambulance team arrives and, checking Aksel's phone, learns his identity, which lets them see his records in the national medical database, to which they have access. Now they instantly know his full medical history, medications, allergies, insurance, and doctor's information. Importantly for Aksel, the data is encrypted, so that anyone else trying to tamper with his phone or ID card would not have that access. This is an example of the power and efficacy of digital identities.

The Estonian model has been so successful that in the 25+ years since the project began, other countries in Europe have followed its lead, also successfully. Author Leslie lived in Finland for more than a decade and has experienced firsthand the joys of living in a society in which information is digitized and centralized as it is in Estonia. When her infant son fell ill while the family was on holiday in northern Finland, the local doctor there had instant access to the baby's medical records before prescribing any new medication. Voting registration is automatic and uncontroversial, and getting an official copy of your birth certificate takes minutes at any magistrate's office anywhere in the country, rather than hours in the grim bowels of a specific Manhattan bureaucratic complex.

Finland has an unusual approach to traffic fines in that the amount you have to pay for a speeding ticket is a percentage of your annual salary. On one summer day when Leslie was stopped for speeding on a highway outside Helsinki, the policewoman was able to pull up Leslie's income tax return from the previous year in her patrol car and tell Leslie right away how much she would owe for her infraction. As Leslie learned the hard way, Finland's approach certainly is an effective deterrent to speeding.

Applying for a school or a new job in countries like Estonia or Finland is very easy. All your transcripts and employment confirmation letters are linked to your ID, with no need to order copies of transcripts or contact previous employers. Think of the time that saves! And if, like Mr. Santos, you claim to have attended Baruch and worked at Goldman Sachs, there are easily available records to prove that your claim is true, or not. A politician's records – whether their tax returns, job, or voting history – are all in the public sphere, making the country a much more transparent place.

According to Transparency International, these public digital ledgers do indeed contribute to a lack of corruption. In the 2022 version of their Corruption Perceptions Index, Finland came in just behind Denmark as the second-least corrupt country in the world, with Estonia just outside the top 10 at #14.[8] The United States took 24th place, just behind the Seychelles. Not bad, but – could be better.

In many ways, the stories of both Representative Santos and Anna Sorokin serve as a cautionary tale for us all, about how a future nation of influencers whose track records are built on their "main character energy" and whose screen presence, charisma, and ability to convert smiles to coins can be part of a long con. What's worse is that, even when caught in the con, these charismatic idols are often allowed to get away with it. They continue to hold their seats in Congress, or in Anna's case, are propelled to fame through book, movie, and network television deals.

We've seen that Virtual Natives tend to locate authority in peers with charisma, rather than in older figures who can point to a lifetime of visible achievement. This is the dark side of that way of thinking: the bad things that can happen when too much trust in surface appearances is not backed up with enough validation. The solution may lie in digitization and authentication.

It's easy to imagine the benefits of having all of one's data centralized in your own wallet and owning that data. Today's Virtual Natives, with their digital literacy, desire for personal agency, and willingness to pick apart existing systems and reject the parts that do not serve their interests, may look to examples outside US borders and realize that this centralized data ownership is what they want. In fact, they may demand it.

Some local entities within the United States, such as Santa Cruz County in California, are already establishing pilot programs for the creation of digital wallets to hold blockchain-based records for government services, beginning tentatively with uncontroversial items like bicycle registration and RV parking permits for its residents.[9] Will other government entities, companies, brands, and celebrities be equally ready to cede total control of the information that they hold? Will they ultimately have a choice?

Changing Expectations: "I should be able to . . ."

We've all been stuck on calls where the company was "experiencing higher than normal call volumes," even if there is no "higher than normal," because that is actually the standard reply. Many companies are complacent, and with messages like this, they are essentially asking clients to wait for poor service. And then once the phone is answered by a human, often low-paid, under-trained staff at an offshore call center, interactions are usually lengthy, cumbersome, and frequently disappointing. Who has time for that?

Virtual Natives, on the other hand, use their phones for text, photo, and video messaging, and very rarely make voice calls. They are far more comfortable interacting with a text-based chat window, or maybe looking for an answer on YouTube, rather than actually talking to a human in real time.[10] "It's just mentally scary to be talking on the phone with a human," says Viktor, 15. "If I'm typing text into a chat window, I can think about what I say before I send it, and I can go do something else while I'm waiting for the reply. It takes much less time that way to get stuff done than it would if I had to wait and talk to a person."

If today's 15-year-olds can already see that traditional customer service methods, such as speaking directly with a call center rep, aren't efficient, and reject them in favor of newer solutions, what will happen when they get older and encounter more complicated processes that *really* waste their time? Many VNs have already created wallets that hold all of their digital goods and tokens, and they'll be rudely surprised to find that there is no equivalent single-database format for storing their driver's license, permits, medical records, and eventually loans, mortgages, and more. Even pioneering Santa Cruz County is creating a separate wallet for bicycle permits, not something that could integrate with a wallet their residents may already have. When VNs discover this lack of innovation, and fragmentation in the few places where innovation occurs, they will start to wonder why. Why *can't* I do this? This is going to create business opportunities for them.

DAOs, or Decentralized Autonomous Organizations, are blockchain-based organizations that, as the name suggests, are not led by any particular person or group, but rather are a legal structure in which all members work equally toward a common goal. DAOs are still in their infancy, but they could soon provide a solution to emerging demands from an economy that is increasingly decentralized, and increasingly virtual. Take car-sharing apps, for example.

By 2030, it's no stretch of the imagination to envision ordering a self-driving car using an app, like Uber. But now imagine that the entire fleet is managed by a collective, or a DAO, whose individual members each own a car. Those owners would contribute to the DAO to maintain the app and pay other marketing and administrative expenses and generate revenues from each ride taken in their own individual cars, which would be out there on the street on their own, ferrying people around without requiring any attention from the member. Sounds pretty easy, right? DAOs are one of the Web3 distributed blockchain technologies that, like centralized wallets, can provide Virtual Natives with the tools to become entrepreneurs and control their assets directly.

How else can we imagine new blockchain-based structures evolving in the future, creating new businesses to eliminate friction that traditional businesses never thought worth addressing? Or is all this new technology too complicated, and it doesn't really matter if businesses evolve? It does indeed matter, and it's driven by culture.

Sparking Joy

In our Estonia example, we can surmise that there was probably a large segment of the population who didn't understand why having internet access was important, and who weren't very enthusiastic about adopting it. History tells us it was a success, but as we know, innovation can be difficult at the actual time when you need to make a switch, especially if you're, say, an older line worker in a pulp and paper plant. Today, embracing online worlds and work processes is difficult to imagine, too. But Virtual Natives are evolving a new set of expectations that are being delivered by technology.

Take TikTok, for example. In 2022, 3.5 billion people downloaded the TikTok app. People turned to the app for news, entertainment, skits, music, and even for local suggestions ("Best things to do in Chicago in 24 hours"). It was the most downloaded app worldwide. Yet this Chinese app, originally called Douyin, was only first released in 2017. How did a simple, unknown app dominate the world's biggest social media market in just five years?

The answer is innovation. The company realized that one of the most popular, and, frankly, pernicious, practices in existing social media was that algorithms tended to share things that got people very emotional. And negative emotions earned social media platforms more money than positive ones.

Former Facebook employee and whistleblower Frances Haugen has explained that the platform prioritized the sharing of "angry, polarizing and divisive content," because an article bombarded with angry emojis actually gets more views – and therefore advertising traction – than one that is well-liked but has fewer reactions.[11] This was not just a problem at Facebook, but common practice across all social apps. Cheery, cheeky TikTok was an island of optimism amidst a sea of social media platforms that took advantage of their audience and focused on harvesting data for the platforms' own ends, rather than giving an informational and emotional reward to its users. No wonder Instagram Reels is modeling itself in TikTok's image.

TikTok is not merely entertaining. It can be informative, joyful, and resourceful. What's more important is that it is highly customized to the viewer. It responds to what people like and what they want, not what makes them angry. To use the terminology of decluttering guru Marie Kondo, when it comes to technology, "Ask yourself if it sparks joy."[12] Virtual Natives expect that technology is there to serve their needs, to give them the tools that allow them to create and share content, brand themselves, and potentially earn money. The Virtual Native expectation of the companies that they do business with is that they put the consumer and creator first, rather than treating them as a data cow to be milked.

Companies need to rise to the changing needs and expectations of Virtual Natives, who will demand increasingly seamless online products and services, and will reward the entities that deliver this positive user experience to them. Just as no one wants to download yet another app, Virtual Natives won't want to bother having to go to one portal to download a health care expense, then open another portal to process the related insurance claim, then import copies of their ID to prove that it's them, and so on. Too many systems today are built around convenience for the institutions they serve, not for the customer base.

It's not just Virtual Natives. According to PwC, "80% of American consumers of all ages say that speed, convenience, knowledgeable help, and friendly service are the most important elements of a positive customer experience."[13] In addition, customers are more likely to try additional services or products from brands that provide superior customer experience, and 63% say "they'd be more open to sharing their data for a product or service they say they truly valued."

What sets Virtual Natives apart from older generations in this desire for excellent treatment is that, according to a different study released by The Center for Generational Kinetics, "this is a generation of pragmatists, self-starters, and entrepreneurs, so if it doesn't exist, Gen Z will build it."[14] These researchers found that more VN members planned to start or possibly start their own business in the future, at 62%. Moreover, the majority of those interested in someday having their own business said that it was likely to be an online business. They're high school students today, but will be businesspeople tomorrow.

So to answer our earlier question about what happens if companies fail to upgrade their systems and catch up with culture, the question to ask is: "Do we want to be Sears?"

Many years ago, before the time of Amazon, there was a company called Sears, Roebuck and Company. Founded in 1892, they published a thick book that they called a catalog. The catalog was up to 1,000 pages long and nearly a foot high, and it was full of pictures of all the things you could order and have sent to your home. Much like today's Amazon, you could find everything from knitting yarn to an entire prefab house, and in many respects the Sears offerings reflected American aspirations and dreams.

Through times of both boom and bust, the Sears catalog was an American staple that could be found in virtually every home. Its products were on nearly every middle-class wedding registry, they helped returning GIs build their first homes, and the book itself was even used as a booster seat in the highchairs of America's babies. Sears opened retail stores across the country. But times changed.

Sears steadily lost its seat at the heart of the American hearth. Big-box stores like Kmart, Best Buy, Target, and IKEA entered the market and Sears found it difficult to differentiate itself from the competition. In 2011, it stopped being profitable and by 2018, after reigning for over a century in business as America's leading retailer, Sears filed for bankruptcy, with 68,000 employees, $7 billion in assets, and $11.3 billion in liabilities. What happened?

The short answer is that Sears stopped innovating, and turned its focus inward. Hedge fund manager Eddie Lampert, who had little experience in retail, took the reins as CEO in 2004. Instead of focusing on customer needs, he focused on profitability and shareholder return. In a single-minded quest for these financial goals, Lampert sold off many of the store's beloved brands and national retail assets, and dramatically reduced staff. He split the company into 30 divisions, each with its own executive team, supply-chain management, and warehousing and distribution system. He announced that Sears would catch up to Amazon (which was, ironically, succeeding by delivering goods directly to customers' homes in a modern-day version of Sears' original business model) by leveraging the internet more effectively, but to no avail. Sears could reorganize all it liked, but the zeitgeist had been captured by more customer-focused shopping experiences, like those of IKEA and Amazon.

The Sears lesson teaches that not all disruption is good disruption. Just rearranging the boxes in the org chart is not the shake-up that a company needs when it is facing changing customer demands.

Increasingly, as Virtual Natives begin to participate in a wider world, they will demand greater digitization of services, and benefits that have recognizable value for them. Companies that respond to these mysteriously changing customer needs with yet another reorg will be missing the point. Successful companies, however, will put rewarding Virtual Natives in ways that they recognize at the center of their strategic marketing plans and corporate growth strategy.

One positive recent example in this domain is Reddit. Their Reddit Community Points (RCP) system, launched in 2020, allows Redditors to own a piece of their favorite communities by rewarding content creators who have, for example, more upvotes on their posts than downvotes. The currency is Reddit's own Moon token, which lives on the Ethereum blockchain and can be swapped for any other token on the Ethereum network.[15] This delivers real benefits for people who actively contribute to a community, directly linked to the amount of the value they created there. And Redditors appreciate the program, which speaks to them in a language they understand.

"Like I literally got 'paid' 164 euros for having fun interacting with this subreddit."

"Wtf lol. Love it."

Reddit has more than 400 million active monthly users. If even half of them participate in RCPs as Reddit expands the program, that's over 200 million holders of Reddit RCPs, which is more than the number of either Bitcoin or Ethereum holders. It is also a rich ecosystem for potential advertisers. Win-win!

Digital transformation is difficult, and usually takes years to achieve, with constant evolution, iteration, and adaptation. Companies have to take on reasonable amounts of risk, while realizing gradual rewards. But innovation and transformation that focuses only on the technology – or the org chart – will be doomed unless it innovates its psychology as well. And understanding the preferences, culture, and skills of Virtual Natives, their desire to virtualize processes and reduce them to their effective essences, in ways that genuinely create value for Virtual Natives, is at the heart of understanding the seismic changes that are headed our way.

Virtual Natives Have the Edge

In this chapter, we've been looking at the need for companies and other institutions to digitize and modernize in order to remain attractive to Virtual Natives in the long term – or prepare to be disrupted by others who do a better job of it.

Virtual Natives know, through years of practice, how to use digital tools. Making money in the Metaverse will be second nature for them, because they're the ones who will be building it. To understand what that means, take the example of Roblox. Roblox attracts about 54 million daily average users, with an average age of 14 years old. Those children are growing up accustomed to building worlds and creating activities in them, and buying and trading in-game assets ranging from skins and accessories to virtual vehicles and home goods. They do this every day.

Virtual Natives are deeply familiar with the market platforms where they buy and sell their items, and are learning from direct, hands-on experience about marketing, advertising, and pricing their wares, forging brand partnerships and sponsorship deals and more. They are savvy self-marketers who know instinctively how to create compelling short-form videos, can deconstruct the success of viral content, and know how to segment and demographic targets.

They won't need linear step-by-step training and onboarding as earlier, less experienced employee cohorts have. Instead, they will need to learn how to integrate corporate skill sets and practices into their own routines, a process that won't work well if there's too much rigidity on the corporate side. Much like coding and gameplay, VNs will have already developed highly personalized workflows, working habits, and routines long before they are hired. Companies that understand VNs, and focus on delivery against deadlines rather than fixed times, places, or prescribed methods of work, are the ones that will best be able to harness their considerable skills. Companies that mistakenly assume that their VN employees' youthful age means that they should be surrounded by guardrails and guidelines won't have VNs on the payroll for long.

Substance and Shine

We began this chapter with the cautionary tales of George Santos and Anna Sorokin, whose escapades highlight the downside of relying too heavily on charisma rather than trust, and then followed the theme of simplified identity and document validation all the way to the need for greater corporate responsiveness through digital innovation, and ended identifying Virtual Natives as the best ones to do the task. *Whew!*

Perhaps the strongest common through-line across all these related topics is that of sparking joy. While it's prudent to take a "trust, but verify" approach to individuals who perhaps seem a little too good to be true (as the saying goes, "trust everyone, but cut the cards"), we know for sure that Virtual Natives are headed for disappointment as they embark on the great life adventure of adulthood, and learn just how much bureaucracy there is to filing a health insurance claim or applying for a mortgage. Happily, the same mechanisms that can help us understand whether people really are who they say they are can also help us create intelligently streamlined systems that will spark joy in the hearts of those who use them, instead of generating the sense of lurking doom that we often feel as we face an interminable bureaucratic process today.

With their life experience of using digital tools to abstract processes down to their essentials, their desire to take action to make things better, their willingness to work with computers as partners, and their overall very high digital literacy, Virtual Natives are the ideal employees to reimagine and implement new processes that will benefit both themselves and the companies that they work for. The trick will be to ignore how old they are (or aren't), and to listen to what they say when they tell us how things can be made better. As content creators and charismatics, they offer both shine – and substance.

17

Main Character Energy

As WE SUGGESTED in the previous chapter, by the time they join the labor pool, Virtual Natives have already formed work habits that they will import into their business lives. But are businesses prepared to welcome these trends?

When we looked at Virtual Native workplace preferences in our discussion of personal agency and control in chapter 8, we focused on how their spirited sense of self-worth, often supported by the direct experience of making their own money digitally before entering the workplace, gives them a wider sense of options than previous generations have had.

Not all companies are ready for this change. Let's look now at some of the key shifts in the perception of authority and agency in the workplace, the largest points of friction, and how businesses can adopt successful strategies to attract and retain those Virtual Natives whose digital skills may make a strategic difference to the company over the next decade.

Working from Home: A Structural Shift

"America, it's time for a nationwide family reunion," reads a 2021 opinion piece in the *New York Post*.[1] "Come back to see the geezer who stands by the urn of Chock Full o' Nuts offering dad jokes; the ambitious young upstart quietly scheming his way into middle management; the girl in the corner who laughs too loudly at her social media . . ."

"It's time to go back to the office."

The global response to the Covid pandemic of sending people home to work was originally a move to reduce the transmission of lethal germs, nothing more. Months away from the workplace turned into years, and the white-collar population of the world had often come to appreciate a workday that didn't begin and end with a long and frustrating commute. It seemed as though office life had vanished overnight. No more "office husband" or "office wife," abandoned birthday cake in the office pantry, lingering by the coffee maker on a Friday morning trying to look sober, or watching the first snowfall and figuring out whose train line would be delayed; no more racing to the bathroom before a long afternoon meeting, during which you'd play jargon bingo, doodle, or text friends from under the desk – all those mundane moments, the forced camaraderie, the awkward and often unnecessary rituals that made up office life and culture were now, for the most part, gone.

Even as global vaccination rates rose and people slowly returned to public spaces, first with masks, then without, commercial vacancy rates remained stubbornly high, creating a serious problem for both real estate firms and city managers whose tax base had precipitously dropped.[2]

But the group of people who seemed to complain the most about their employees' unwillingness to trade their pajamas for a suit was that of CEOs around the world. In one example, British billionaire and businessman James Dyson responded to UK plans to give all workers, even those with no seniority, the right to request flexible working arrangements, by calling the arrangement "staggeringly self-defeating."[3]

Phones 4u founder and billionaire, John Caudwell, fulminated in 2022 against what he saw as employees' "growing sense of entitlement," and their seeming belief that jobs "exist for their own convenience rather than to serve customers or the public."[4]

And Jamie Dimon, CEO of JP Morgan Chase, bluntly stated that working from home "doesn't work for people who want to hustle," implying that those who stay home are less productive and less interested in their careers.[5]

"When you really peel it away, work from home, . . . and all the inefficiency it brings, is in my mind a very slow cancer that is very silent but growing on this economy," said Marc Holliday, CEO of SL Green, New York City's largest commercial real estate landlord.[6]

But . . . are these CEOs right? We're not sure who's going to break it to them, but research has consistently shown that remote work can be just as effective as, if not more so than working in an office.

A two-year study of three million workers at 715 US companies, including many from the Fortune 500 list, revealed surprising results: Working from home improved employee productivity by an average of 6%. It seems that instead of encouraging slacker behavior and letting employees get away with sleeping until noon, remote work makes employees feel they have greater control over their work, which makes them happier and more motivated.[7] Remote work can also lead to cost savings for both employees and employers, as it reduces the need for commuting and office space. And employees appreciate having more control and more time in their lives.

Even as corporate executives at firms like Netflix and Goldman Sachs continue to insist that they "don't see any positives" about allowing their staff to work from home,[8] their workers feel differently. In a 2022 survey report entitled "Americans Are Embracing Flexible Work—and They Want More of It," McKinsey revealed that a massive 87% of employees who were offered the opportunity selected the option to perform at least some of their work remotely.[9] A survey of 3,000 people working at companies including Microsoft, Meta, Cisco, Google, and Amazon found that 64% said they would choose a permanent work-from-home option over a $30,000 pay raise.[10] Time is money, after all.

In the comments section of a *New York Times* article about the office perks offered by some companies to get workers back behind their desks, one reader captured the pushback expressed in many of the other comments as well when she wrote, "Ridiculous. The only people agitating to get back to an office are executives desperate to justify their leases and middle managers desperate to justify their existence."[11] Another fan of remote work sarcastically commented, "All these imaginary inefficiencies have allowed me to produce 500% more work at home."[12]

A study of teleworkers conducted by Stanford University had even starker conclusions about the benefits of working remotely.

Researchers there found that compared to their in-office counterparts, remote workers were a full 13% more productive. In fact, they noted that teleworkers logged 15% more calls overall, 4% from more calls per hour and 11% from more hours worked. This increase in productivity may be because, in many cases, workers at home have fewer distractions and interruptions than in a traditional office setting, and of course they no longer have that commute to deal with.[13]

Virtual Natives often have yet another incentive for working at home instead of going into an office every day. In 2023, Instagram found that nearly two-thirds of Gen Z social media users plan to start some sort of project in which they use social media as a tool to make money.[14] Calling "being a creator the new part-time job," the study confirmed the tendency toward having a side hustle that we mentioned as common among VNs in chapter 8. So many young workers have developed lucrative second jobs during their time at home during the pandemic that some companies, such as Japanese trading firm Mitsui, have removed the ban that they once had on their employees' out-of-hours entrepreneurship, freeing up their VNs to "become YouTubers or startup founders."[15]

In fact, side hustles may be just the minimum that working from home allows. A 2023 article in *Fortune* magazine discussed anecdotal evidence that the rise of Generative AI tools like ChatGPT was helping knowledge workers rip through their work at such a lightning speed that they felt able to take care of two and even three full-time jobs at once. According to their sources, one person interviewed confidentially "has been using ChatGPT to do two jobs and is hoping to add a third, increasing his compensation from $500,000 to $800,000. He considers himself part of the FIRE movement ("Financial Independence, Retire Early") and is not yet 30."[16] So much for the "lack of hustle" that JP Morgan Chase's Jamie Dimon assumes home workers suffer from.

And in another angle, it turns out that working from home can be much more pleasant for members of minority groups.

A survey conducted by Future Forum, the Slack think tank, found that 81% of Black respondents and 86% of Hispanic respondents preferred a fully remote or hybrid workplace, compared to 75% of white respondents.[17]

"I actually like not having to go into the office and be constantly reminded that I'm the only Black woman there," said Courtney McCluney, an assistant professor of organizational behavior at Cornell's ILR School. McCluny and others cited relief from being subjected to a constant battery of microaggressions regarding their hair, food, or cultural references as reasons why they'd prefer not to return to a physical office.[18]

Remote work can be a welcome relief to employees who want to avoid superficial interactions and the need to uphold office norms with which they fundamentally disagree. For Virtual Natives, the most racially and ethnically diverse generation in America history, with 48% of them identifying as non-white, this may be especially true.

The bottom line is that this change may be more or less permanent. In the twentieth century, improving agricultural technology freed millions of young people from needing to work on the family farm, and allowed them to move to cities to pursue industrial or office jobs. Later, electricity enabled longer workdays and night shifts. Today, our vastly improved connectivity and computing infrastructure is enabling the next big shift around where people are physically located. If your connection is fast enough, and secure enough, and you're able to use digital tools to remotely collaborate with your coworkers, then why insist on an unproductive return to an earlier format just for the sake of it?

Greed Is Good and Closers Get Coffee

On a conference call with JP Morgan Chase investors, CEO Jamie Dimon was again pressed to comment about extended work-from-home allowances and its effect on corporate culture. He snapped. "We want people back to work . . . and everyone is going to be happy with it, and yes, the commute, you know people don't like commuting, but so what?"[19]

To understand some of what shapes this rather common (and rather absolutist) management reaction to the concept of remote work, and the schism that it is widening between them and Virtual Natives, let's go back to the time when today's CEOs were just starting out their own careers. Who were their role models and what was office life like for them?

From the 1900s through to the dot-com era in the early 2000s, people wore suits to work. They were encouraged to wear certain colors, like navy blue, or black, with crisp white shirts. Visually, office workers were relatively interchangeable.

The power suit hit its peak in the 1987 film *Wall Street*. In the movie, the young protagonist looks up to the lean, square-jawed Gordon Gekko, a symbol of sartorial sleekness. Gekko's wardrobe consisted of coordinated ties, mouchoirs, socks, accenting pinstriped designer suits from Italian menswear purveyors like Armani and Gianfranco Ferre. The look served as inspiration to a generation of men. But it didn't stop at the suits. The film's core theme was that bald-faced capitalism is good for society.

"Greed," Gekko stated, "captures the essence of the evolutionary spirit. . . [It has] marked the upward surge of all mankind." Then came the phrase which has since been included in the American Film Institute's list of the 100 greatest film quotes of all time: "Greed, for lack of a better word, is good."[20] This was great stuff. Young men in cities across the country quoted him as they dressed themselves in killer suits and sharpened their metaphorical knives in corporate boardrooms. Gordon Gekko said that greed was good, and that was all they needed to know.

In another film that captured the sentiment of the time, the 1992 movie *Glengarry Glen Ross* focused on a group of real estate salesmen who were so desperate to hold onto their jobs that they each decided they would do nearly *anything* to keep them. The film held a mirror to a system in which corruption and predatory practices were tolerated, even encouraged, if that was what it took to win. Anyone who held back, or refused to play the game, was a failure. The film's most famous line, "Coffee is for closers," delivered with a steely look designed to melt the weak, became a catchphrase.[21] If you haven't closed a deal recently, you don't get coffee, or cash, or even a ride in the boss's car, because you haven't earned it, you loser.

For men and women in the formative period of their careers in the 1980s and 1990s, this was all just normal work culture. And this is the context for the frustrated reactions of the CEOs who can't stand to see their workers openly revealing that maybe work isn't the most important thing in their lives. Even though efficiency is up, and "closing" is therefore happening, traditional executives are getting the disturbing sense that their employees are no longer motivated by the greed and competition that drove them. And they're right.

Today, younger generations don't feel the urge to burn themselves out over a career, because they have seen in the examples of their parents and older siblings that neither jobs nor employers will love you back. They have seen CEO pay balloon to over 300% of employee pay, while the salary of average working staff has held constant since the 1970s. Virtual Natives are rethinking what success means for them in a system that they have not designed, and which may not fit their personal and emotional needs.

The *Harvard Business Review* reports that since 2005, "the importance of meaningfulness in driving job selection has grown steadily," and that today, 9 out of 10 workers would prefer to have a meaningful job, even if they made less money.[22]

How will these Virtual Native desires align with the employment structures and hierarchies that have already been constructed? Is it just a matter of insisting that younger generations get over themselves and conform, as their elders did before them?

The Uniform Versus Uniqua

In the army, one of the first things they do when you join is shave your hair. One official reason given for the practice is the need for "field sanitation," that is, to deter the spread of hair and body lice. Additionally, a shaved head reduces the likelihood that the enemy can grab the conscript by the hair and, for example, slash their throat. But one of the most fundamental reasons that the army cuts your hair is to address – and reduce – your ego.

Cutting your hair is to detach you from what is arguably the most important and personal signifier about your appearance, other than the color of your skin. A long hairstyle can tell the world that you're relaxed and easygoing, while a short one can suggest that you're energetic and possibly conservative. Long sideburns indicate a retro man, a mullet suggests a country man, and so on. In the biblical tale of Samson and Delilah, Delilah discovers that the source of Samson's strength is his hair. By cutting his hair, Delilah expresses her psychological domination of Samson, in much the same way that armies the world over exercise their right to control the appearance of their conscripted.

After having their hair cut, army conscripts are given a uniform. The word "uniform" is from the French for "one look," and indicates that the purpose is indeed for all soldiers to look the same. The conscripts now form a single body, and no individual stands out from the sea of sameness. From there, a battery of grueling physical exercises, chores, and field duties breaks down whatever sense of personal freedom the recruits came into the army with, before building them up again as a member of the unit, a collective person, someone who puts the team before the group. Social scientists have proven that the old adage that "what doesn't kill you makes you stronger" may actually be true, and it can, in fact, be an aid in forging relationships among people who went through the experience together, which is of course why soldiers are put through those tasks in the first place. Whether plunging one's hand in ice-cold water, or eating hot chili peppers, studies conducted by the University of New South Wales found that pain experienced as part of a group was "a particularly powerful ingredient in producing bonding and cooperation between those who share those painful experiences."[23]

Many senior executives will suggest that the formative training and difficulties they experienced as part of their work life are a righteous path to success. If I could play the game, wear the uniform, make sacrifices, follow the rules, and I thrived, the thinking goes, so shall others follow.

Virtual Natives are coming from a completely different place. They live their lives out loud, spotlighting themselves and others on social media. They're used to highlighting their uniqueness across multiple platforms online, and to accepting and celebrating the individual differences of others. The worst insults that VNs have for each other are terms like "Basic" or "NPC," that is, being a non-playable background character in a game. Being just average, a blurry face in the crowd, is highly undesirable – you have to stand out in some way. In the Nickelodeon children's cartoon *The Backyardigans*, which ran from 2004 to 2013, the central character, the only one to appear in every episode, was named Uniqua, a take on the word "unique."[24] In a sense, all Virtual Natives are named Uniqua, and they're not ready or willing to drop their glowing individuality to become subsumed into a hierarchical corporate culture in which the most that is expected of them is that they toe the line.

Valuing the Human Beyond the Resource

Much has changed in the years since the birth of the internet and personal computers. So, too, should management be thinking about change. Even as companies consider digitalization to improve their relationship with their customers, as discussed in the previous chapter, they should also leverage the increasing virtualization of the workplace and its functions as an opportunity to reimagine corporate culture and employee engagement.

First comes the right mindset. Humility helps. Professor Ed Schein of MIT, who literally wrote the book *Corporate Culture*, told *Forbes* magazine that "leaders have to become much more humble and learn how to seek help, because [their] subordinates . . . will be much more knowledgeable than they."[25] This is a far cry from the sneering "no coffee for you, loser" attitude of superiors toward their subordinates in *Glengarry Glen Ross* days.

Rebuilding with the Employee at the Center

A perhaps odd example of leadership humility comes from Steve Jobs, who was famous for his autocratic leadership style, coupled with a meticulous and exacting eye for detail, demanding that others apply the same uncompromising, intellectual rigor and emotional investment into their work. Jobs felt strongly about appointing and promoting executives who could outshine him in some way, saying, "It doesn't make sense to hire smart people and tell them what to do. We hire smart people so they can tell us what to do."[26] Even the supremely confident and capable Steve Jobs was smart enough to be humble at the right time.

Second, employers need to honor the lives that Virtual Natives have outside of the office. As Matsui has recognized by allowing side hustles, Virtual Natives often have not only a social media presence, but very often also have their own external brand, which they will insist on retaining and holding separate from their corporate identity. VNs have multiple social accounts and well-established networks across platforms where they are accustomed to expressing their opinions. If companies are sincere about respecting independence, talent, and diversity, they will have to give them the autonomy necessary for reasonable self-expression and social presence outside the workplace. This autonomy also means that the VNs are not loyal to their employers, but to themselves, exactly matching the amount of loyalty that corporations show their employees. If a VN is holding two full-time jobs simultaneously, and is actually delivering the work that is expected of them in both, should the employer even care?

Third, and most important, is simply to treat VNs as human beings, and to recognize that they are Uniquas, instead of uniforms. The website Great Place to Work has found that the real difference between *good* and *great* companies is when leadership acknowledges their employees' humanity.[27] They stress the importance of seeing employees "as whole people, with family, hobbies, and passions that they bring to work each day. When relationships are strong, employees feel energized and bring their skills to the table to collaborate on organizational goals."

The most successful leadership style with Virtual Natives is not to be paternalistic, authoritative, or even visionary, which suggests a single leader who has all the answers and needs little input from their team. Instead, the leadership style with the best fit for Virtual Natives is most likely to be *empathetic*, and even *trusting*. Today's leaders of the young need to respect the autonomy of a distributed workforce, lead by example rather than decree, hire smart people who are motivated to reach stretch goals beyond current industry limitations, and trust them to figure out how to get there. At the same time, empathetic leaders will acknowledge and respect their employees as humans who bring their whole selves to work, and who have a fully developed world beyond the office, even if that office means they're working from home. Leaders of Virtual Natives will also understand that charisma and especially authenticity are what will encourage their employees to respect them and their authority; merely asserting that they should be respected because they're in charge won't get managers any respect at all.

The good news is that when in a system properly aligned to their way of working, Virtual Natives are ideally productive employees. Natural freelancers, self-managing Virtual Natives fully expect to juggle tasks simultaneously across different projects, each with different systems, teams, goals, and timelines. They are good at generating content, selling, and storytelling. They are the main characters in their own lives; they work independently and asynchronously, against deadlines, not time clocks. Expecting them to sit at a particular desk in a particular office from 9 a.m. to

5 p.m. will only puzzle them; instead, telling them that they and their team have a deliverable due on Thursday at 5 p.m., and they can work on it how, when, and where they wish, will get the desired results.

With their niche skills honed over time, working with friends and AI tools online, and relying on YouTube to fill in any gaps, there may be little in the way of technical training that VNs need. They are more likely to require time for iteration and feedback cycles, along with access to open communication with leadership. To retain them, companies will need to demonstrate that their VN employees are valued beyond their mere ability to contribute to the bottom line by being "closers." Instead, leaders will need to nurture the human beyond the resource by inspiring their individualism, supporting their wellness, demonstrating a respect for diversity, and identifying a path for advancement and remuneration – that is, clearly communicating how their VNs can "level up" in their careers.

18

Future Forward

Virtual Natives and the New Tech Landscape

In Part One of this book, we identified the Virtual Natives and what defines them. What we have found is that Virtual Natives are self-driven, creative, impatient people, casually expert in digital everything, always ready to rethink and replace mechanisms in the modern world that don't make sense, especially traditional structures that may no longer serve their original purpose well. They have been trained by games to have agency, and by Instagram and YouTube to accept and appreciate people from all walks of life. They experiment with both their own identities and how much they can accomplish with artificial intelligence as a sidekick. Because of the reach of their voices and opinions, thanks to digital platforms, and their ability to reshape things through the use of digital tools, Virtual Natives have the potential to be the most powerful and influential generation of young people that our world has seen.

In Part Two, we looked at how the Virtual Native mindset is likely to respond as this generation enters the adult world, and how their expectations and familiarity with digital tools will create both tensions and new solutions as they encounter non-optimized structures that older folks have gotten used to just shrugging and accepting. This is where the magic will happen, for those bold enough to listen to what this younger cohort has to say, and their ideas for solving problems from their digital mindset. Business opportunities will abound as Virtual Natives find more efficient ways to structure their lives, and disruptions will necessarily also abound in their wake. Companies that can think flexibly and move beyond "the way things have always been done" to embrace a Virtual Native mindset have the greatest potential to lead these disruptions, rather than becoming the dinosaurs who are wiped out by the Virtual Native meteorite.

In this final chapter, we're distilling our analysis of Virtual Natives into a summary of how we see the future will evolve in several important areas. We're painting with a broad brush, aiming to be evocative rather than prescriptive, and using everything we've discussed up to this point as the colors on our palette.

The Future of Labor: Radical Self-Reliance

For centuries, we've lived with the assumption that people need to go out and get a job. Any job, really, so long as you can pay the bills. In that sense, the power of labor relations has long been understood to be in the hands of the employer, who has the money to pay employees, who in turn need money to live. But culture has shifted, led by Virtual Natives who increasingly see ways to make money that involve traditional employers only minimally, or not at all.

Side hustles have become an increasingly important part of our lives, and to younger generations, entrepreneurialism comes naturally. They wield the world's most powerful technological tools and have an innate understanding of their function and potential. Virtual Natives were born after social media had already begun to flourish and have grown up shaping it. Importantly, they have moved far beyond the neighborhood lemonade stand and know how to monetize their skills better than practically every generation before them. Children under the age of 15 are contributing to household earnings, some even earning six and seven figures annually from their work. These children know how to build and market their own products, and are making important daily decisions about how to earn, invest, and spend their own money. They develop these skills and have these credentials long before they enter the ranks of traditional employment, which employers must be mindful of.

Take Sheila as an example. At 28, she got a gig job at a global tech firm. She was also very active in a global tech association, the VR/AR Association, which gave her expertise that was part of what made her valuable to the firm. But when told that she would have to put her social media activities on hold during her tenure at the company, her immediate response was, "No way!"

"I've worked really hard on my brand. If they don't like it, they can stuff it."

Virtual Natives are the owners of their own brands and assets, and generate both active and passive income online. They create their own content, and evolve their skills online alongside their peers. They take considerable pride in their work and expect to be paid equitably for their labor. And they have access to tools and a support network as never before.

Virtual Natives are taking control of their future income without needing to depend on companies to provide for them. Knowing that their side hustles will help sustain them in lean times, they won't be afraid to walk away from pointless, unfulfilling jobs, while posting their dramatic exits on social media. They will also not be willing to put up with "coffee is for closers" corporate toxicity. Values matter. All of these factors are driving an evolving relationship between Virtual Natives and the companies that engage them. Work is increasingly at-will for both employer and employee and, increasingly, on-demand.

The Future of Education: Education as a Service

Education as we know it today is ill-suited to meet the needs of this new generation who are themselves beginning to question the extrinsic value of a costly college education, with 90% of 2023 Instagram survey respondents citing real-world experience as being more valuable.

There is no question that AI will radically alter the shape of how courses are taught and homework assigned. The traditional premium on original content may give way to new forms of expression that incorporate machine-generated content, rather than fighting against it. This may help an entire new industry to form.

Working from home will require individuals to be more proactive about their own career development. Virtual Natives will continually upskill themselves in their particular trade or craft according to their needs, using resources such as YouTube, Udemy, or Coursera. Companies need to be aware of this and focus on providing enterprise-specific internal mentorship and coaching, while rewarding those who do pursue external education opportunities with tokens, cash, or other incentives, rather than ignoring their employees' outside education activities altogether, as often happens today. Modular education, spearheaded by the employee, will fuel the overall trend toward education-as-a-service. Expect education to be a lifelong endeavor.

The Future of Collecting: "Show Me What You Own and I'll Tell You Who You Are"

Collecting objects is a way of communicating your values, status, interests, and erudition to the rest of the world, and it is as old as time. Ancient Romans collected Greek statues and Egyptian obelisks. The Medici commissioned artists to paint on the very walls of their homes. Young European aristocrats in the seventeenth and eighteenth centuries traveled the world to collect artwork and curios which, once back home, would be stored in display cases to be admired by visitors. These collections, whether of art or odd ephemera, were as much a testament to their owners' wealth as their good taste.

In the modern era, creating and collecting digital assets is a thing. Is having a collection of digital images of dogs, apes, punks, or squiggles really so different from having a collection of Barbies, or a stack of yellowing sports cards? For some, the notion of collecting digital goods with no tangible physical or materiality is hard to conceive. For Virtual Natives, for whom both digital and physical goods have reality and validity, the value of NFTs and other digital assets is obvious.

Show me your wallet and I'll tell you who you are: Like gathering the objects in early curiosity cabinets, amassing digital collectibles is a pursuit that reveals the collector's aesthetic values, intellectual pursuits, personal interests, wealth, and social standing. Collecting is the manifestation of a very fundamental human desire to surround yourself with things that represent who you are, or would like to be. The nature of the objects and where they are stored is, in the end, irrelevant. The Virtual Native curio cabinet is a wallet, whose contents signify the wealth, taste, erudition, and sense of humor of its owner.

The Future of Social Life: It's Global

For Virtual Natives, social life is active, and there is no difference at all between in-person friendships in the physical world and those maintained across a variety of apps or virtual spaces. VNs' social lives have evolved into a technological ecosystem formed around the individual and increasingly customized specifically to them. VNs are each their own main character, the hero of their own stories, and the star of their feed. The way they use their public platforms is not as vanity-driven as the activity on early social media; rather, VNs use them to create and foster deeper meaning and connections. They do value face-to-face interactions and Facetime video calls, but when it comes to really hard problems, they may prefer to talk it out via avatar.

Virtual Natives were born into a world where cosplay and SantaCon are already part of popular culture. They grew up seeing "no pants day" celebrated on subway cars around the world. They regularly inhabit new characters, as male, female, animal, vegetable, or fantasy creatures in games where they forge new identities and connections. Gone are the days when the only available pool of possible friends for the average high school student was the other kids at the school, with the possibility of ostracism for those who march to the beat of a different drum. No longer limited in their social choices only to the people that happen to live nearby, VNs can now connect with people who share their interests across a host of new media and games. Their relationships are no longer regional, but interest-based. Thanks to translation algorithms and other digital tools, language is fast disappearing as an obstacle, and VNs are able to communicate with people from all around the world. They learn new languages, cultures, and cuisines on a daily basis. Their perspective is a global one, and their exposure to other ways of being and doing makes them more worldly, connected, and hopefully, more compassionate.

The Future of Entertainment: Interactive and Revelatory

It has been said that you can discover more about a person in an hour of play than in a year of conversation. Play is a way in which children act out social scenarios, test their skills, prove themselves, and explore ways of being in the world as independent actors. In games, children learn to work together and learn about strategy, execution, leadership, decision-making, and the nature of trust, all through direct experience. This happens as fluidly online as in the physical world; in fact, the level of challenge and requirements of strategic cooperation are often far greater in games than they are in, say, the classroom.

Online, while children are in regular contact with other children of their same age, they are also the main character of their own experience, which can be customized for them. From *Poké-mon Go*–like treasure hunts in their physical home and neighborhood, to online battles and challenges, they have access to a wealth of game worlds, story worlds, and new experiences, each full of ritual and lore. As Bruno Bettelheim wrote of fairytales in *The Uses of Enchantment*, "The child intuitively comprehends that although these stories are *unreal*, they are not *untrue*."[1] Virtual Natives are able to discover much truth about the actual world and the people in it as they actively navigate through a wide variety of unreal worlds. They may not get their understanding of the human condition from books and literature, but their exposure to other worlds and ways of thinking in online situations provides them with a different source of understanding that is equally valid.

Acting out and exploring our personality, interests, and identities in virtual spaces gives us a new way of storytelling, and for children, a chance to confront and understand the world, and in the process, themselves.

The Future of Money: Digitized and Accessible

For many Virtual Natives, "money" only exists digitally. They don't carry wallets, but instead buy and sell items in both the physical and virtual worlds with credits held on their phone, a piece of plastic, or in their game world.

Decentralized finance and financial tech services are rapidly opening access, participation, and opportunity for Virtual Natives in the new digital economy.

Even at a young age, Virtual Natives are able to open an account on a gaming platform, link it to a digital wallet, earn online, and store their assets in a virtual bank. With age limitations on the ability to hold a bank account in the physical world, asset storage and management on gaming platforms is often the first kind of banking they will know. From there, it is easy for them to move to virtual finance platforms as they get older. With no need to visit a brick-and-mortar office, no loan interviews, and reasonable credit fees, virtual finance is a very attractive option to Virtual Natives who are suspicious of legacy systems and desire control over their financial health. Rather than visiting a physical bank, meeting onerous requirements for documentation, and keeping a minimum balance, VNs have embraced services like Venmo, PayPal, CashApp, and Zelle, which have made it easier for people to pay and receive payment for services. For Virtual Natives eager to use digital tools to ditch cumbersome and unnecessary processes from the past, this *is* banking.

We're All Natives Now

There are many ways in which the Virtual Native preference for using digital tools to virtualize and improve processes will bring new opportunity and profound change to the way our world works today; here we've just looked at a few. The key for those of us who are not Virtual Natives is to see what they are doing and learn from them, despite their young age. Virtual Natives may still mostly be children, but thanks to the digital platforms that they have made their home since birth, they have a far wider base of real-world experience in everything from making money to understanding their own agency than any previous generation has ever had at the same age, and that's why they are more likely to change us than we are to change them.

Virtual Natives are natural disruptors, and both companies and whole industries will rise and fall as a result of their actions and interests. Those who can leverage the Virtual Native mind-set, in part by being flexible enough to attract and retain independent-minded Virtual Native workers, will be the ones to shape our new future as it develops. Those who don't, can't, or won't learn by looking at the world through a Virtual Native lens will find themselves disrupted, irrelevant, and, in many cases, wondering what happened.

Much like early explorers, we are now embarking on a grand expedition. The fact that you made it this far shows that you, too, are seeking to evolve your understanding of the new landscape and likely already belong to the cohort of people pushing the boundaries of what is possible and what is yet to come.

As the good citizens of Estonia will likely attest, a hard pivot into a technologically forward society is not an easy task. It will require commitment, vision, and conviction.

As we've talked about Virtual Natives all through this book, we haven't been too rigid about the age ranges, because while there's a big Venn diagram overlap between Virtual Natives and Gen Z and Gen Alpha, it's not an exact fit. There are plenty of people with an open, inquisitive, digitally oriented mindset above the age of 30 who are indeed true Virtual Natives. It will take Virtual Natives of all ages to drive the cultural revolution – are you one of them?

Like age and, as we've seen, money, being a Virtual Native is notional. It is a fluency and a state of mind. For people born into this era, it is natural to them, but that doesn't mean that you can't learn or adopt or embrace the promise of this exciting new era. With the tools and framework set out in this book, you are already well on your way.

Welcome to the club.

Epilogue:
Invisible Architecture

ONE KEY ENABLER of just about everything that Virtual Natives do that sets them apart from previous generations is infrastructure. Without electricity, without connectivity, without software, hardware, and artificial intelligence, Virtual Natives are just like the rest of us.

We tend to take our amazing infrastructure for granted, but none of the future that we currently envision will come to pass without continuing significant investment from the private companies and utilities that create the digital tools upon which we all depend.

Infrastructure is, and always has been, the "invisible architecture," the vital foundation upon which we build our lives, and the means by which we continue to improve them.

An Enduring Miracle

At the surface, looking with the naked eye, one can only make out a pit in the arid earth. Were it not for a series of orderly holes of the same dimension, one might easily overlook them. Yet these holes, which dot the landscapes of the Middle East and Central Asia, are a sign of one of the greatest early feats of engineering known to man: the Qanats of Iran.

The gardens of ancient Persia were, and remain, one of the world's greatest wonders. Arriving from a distance across a harsh and unforgiving desert, travelers of the past would be astonished to see what seemed to be a mirage in the distance, but it was no mirage: Almost miraculously, the Persians were able to coax palm trees, fragrant tuberose, crocus, peonies, pomegranate, lemons, and fig trees from the reluctant desert.

"Wherever the Persian king Cyrus resides, or whatever place he visits in his dominions, he takes care that the paradises shall be filled with every thing, both beautiful and useful, the soil can produce."[1]

But this is really a tale of infrastructure. What enabled the magnificent desert palaces of Cyrus and his successors, and their lush gardens, was a radically effective technology called the "qanat."

Qanats originated in ancient Iran, one of the driest regions of the world, in approximately 3,000 BCE. They are gently sloping tunnels that conduct water along underground channels by gravity, often over many kilometers. They bring water to arid places, allowing both plants and people to flourish in what would otherwise be dusty and desolate conditions. It was thanks to the invention of the qanat that the ancient city of Persepolis, capital of the Persian empire, came to be regarded by many as the most luxurious city in the world, famed for its opulent palaces and, of course, its exquisite gardens.

The qanats were a marvel of engineering that opened up resources and allowed economies and trade to flourish. They had a major influence on village socioeconomic development and organization, ownership, and land tenure agreements. Qanats were the invisible engine powering the economic development and the ultimate expansion of the Persian Empire. Yet, like so much infrastructure today, they were underground and therefore invisible. You had to know they were there to, well, know they were there.

Hidden architecture, the infrastructure that remains invisible to the naked eye, has been holding society together for millennia. It allows people, plants, and systems to flourish in both unexpected and faraway places. It is also what we, as humans and consumers, largely take for granted. The people who lived in the palaces of Persepolis may not have understood or appreciated how many qanats underground made it possible to plant their beloved gardens and sustain a population. They could focus on the buildings on top, because of the presence of the engineering below.

Computer scientist Alan Kay, who won the 2003 Turing Award, once quipped that "technology is anything that was invented after you were born."[2] Similarly, we can almost define "infrastructure" as anything that you depend on but never think about, like a qanat – or the internet.

The Digital Divide

Virtual Natives are the generation born from about 2005, who grew up with the iPhone, and used apps as their first experience with computing. Their first car ride could well have been in an Uber. By the time they were preteens, Snapchat's augmented reality lenses were altering the pictures they took of their friends, and they could search for digital Pokémon on the streets of their neighborhood with *Pokémon Go*.

Virtual Natives weren't born into the internet – they were born into the Matrix, a complex interconnected set of tools, technologies, graphics chips, hardware, software, batteries, cloud storage and payment systems, all designed to make our lives richer and more seamless, connected, intuitive, productive, healthy, entertaining, and educational. VNs live a large part of their lives in fully immersive 3D worlds with sophisticated narrative arcs and endless branching story lines. This generation is creating sophisticated short films on handheld technologies that would have been a dream for a young Martin Scorsese or Francis Ford Coppola.

But we can't leave the topic of Virtual Natives without acknowledging a hard truth. There are plenty of young people in the United States and elsewhere who would love to be Virtual Natives, if it were not for one thing: the digital divide.

In 2020 two little girls in matching pink hoodies in Salinas, California, sat cross-legged, huddled together on the cold cement sidewalk next to a Taco Bell. The reason they were staying outside wasn't because they didn't love Taco Bell; they did. The problem was that they couldn't afford to eat there. The reason they were there was the internet. They needed Wi-Fi access for school. But like 40% of Latinos in the state of California, they didn't have internet access at home, and without access to Taco Bell's Wi-Fi, they couldn't even attend class online.

A shocking 42 million people in America today, about 8% of the current population, have no access to high-speed internet.[3] That number rises to 25% in rural areas. The FCC defines broadband as having download speeds of a minimum of 25 megabits per second (or Mbps) and upload speeds of at least 3 Mbps. Currently, the US government is actively working to fix this undeniable problem, but given the challenges of geography, it's going to be a hard gap to close.

The internet is a lifeline that gives people access to education, health care, jobs, entertainment, and community and emergency services. Without it, people can slip through the cracks, unable to advance economically or socially. Having access to the internet delivers significant benefits to society, from improved digital literacy and digital skills to providing greater opportunity for economic equality and overall economic growth.

The metaverse is a fast-growing space that promises many future opportunities for Virtual Natives who have access to it and can take an active role in shaping whatever it becomes. At the same time, the democratization of computing through natural-language interfaces is reducing the need to learn to code in order to get a computer to understand you, hinting at great and even unimaginable things to come as concepts like the metaverse and language-driven image creation are braided together in new ways by imaginative Virtual Natives.

But we must never forget that having regular access to hardware and basic internet is a prerequisite for success.

In addition to the worrying lack of universal access that exists in the United States today, education systems are often a problem as well. Led by older educators, they are frequently rooted in the past, unaware of how the world has changed and that what they should be doing to prepare their students for the future has changed as well. In a recent international survey by Dell Technologies, 37% of Gen Z respondents said that their education was not preparing them with the digital skills they needed to build a career. In fact, 56% said that their official education included either "very basic to no digital skills education at all."[4] Virtual Natives often get their digital literacy training from parents or siblings at home, or from friends, or from YouTube. Not from school.

As we continue to give shape to the metaverse and the other digital domains and tools that make Virtual Natives what they are, it is our collective responsibility to craft a "digital [and educational] environment that is truly inclusive and accessible" – one "that gives everyone . . . the chance to create and participate."[5]

Today, there are still over 30,000 qanats in Iran. They are part of an engineering marvel that laid the foundations of an empire. The invisible architecture of the ancient Persian empire created vast networks of underground canals and ancient aqueducts that defied gravity, harsh sunlight, and distance to distribute the valuable resources that allowed economies and trade to flourish. Virtual technologies will likewise open up a host of economic opportunities to trade across millions of new virtual worlds, augmented realities, games, and platforms for VNs to create, share, and explore.

As Virtual Natives prepare to forge the path, it is our aim to successfully pass the torch, so they might light the long road ahead for us all.

Notes

Preface

i. Marc Prensky, "Digital Natives, Digital Immigrants," *On the Horizon*, MCB University Press 9, No. 5 (October 2001). http://www.marcprensky.com/writing/Prensky%20-%20Digital%20Natives,%20Digital%20Immigrants%20-%20Part1.pdf.

ii. "Alvin Toffler, Author of *Future Shock*, Dies Aged 87," *Guardian* (February 22, 2018). https://www.theguardian.com/books/2016/jun/30/alvin-toffler-author-of-future-shock-dies-aged-87.

Chapter 1

1. UserExperiencesWorks, 2012, "A Magazine Is an iPad That Does Not Work.m4v," video, YouTube, https://www.youtube.com/watch?v=aXV-yaFmQNk.

2. Genevieve Shaw Brown, "After Gen Z, Meet Gen Alpha. What to Know about the Generation Born 2010 to Today," *ABC News* (February 17, 2020). https://abcnews.go.com/GMA/Family/gen-meet-gen-alpha-generation-born-2010-today/story?id=68971965.

3. "Definition of Virtual," *Merriam-Webster Dictionary* (accessed on April 25, 2023). https://www.merriam-webster.com/dictionary/virtual?src=search-dict-box.

4. "The Real Conversations Podcast by Nokia: Gen Z's Wake-Up Call for the Telecom Industry," n.d. https://futurithmic.libsyn.com/gen-zs-wake-up-call-for-the-telecom-industry.

5. Mya Jaradat, "What Being Religious Means to Gen Z," *Deseret News* (September 15, 2020). https://www.deseret.com/indepth/2020/9/13/21428404/gen-z-religion-spirituality-social-justice-black-lives-matter-parents-family-pandemic.

6. Niklas Melcher, "Metaverse Gamers: Demographics, Playing and Spending Behavior," Newzoo, January 19, 2022. https://newzoo.com/insights/articles/deep-dive-metaverse-gamers-data-on-metaverse-demographics-socializing-playing-spending-2/?utm_medium=email&_hsenc=p2ANqtz-_5o3lnl_iXiNmWhh1P5iDaxhW3aztZETIiteb6aq7hEfTFcTrn8xHw-Q36UocezNMR-ak_EbKm3ZegVIiyn-3Ly_E8JNX5G0auLCWhLk810Bz_q04&_hsmi=201780763&utm_content=201782013&utm_source=hs_email&hsCtaTracking=3d2d8102-2fd9-449c-a5c7-cbb1720bf411%7Cd0015075-b375-4586-bf59-e14840ed3425

7. Carolyn Jones, "Social Media Has Some Significant Upsides for Teenage Girls, Survey Finds," EdSource, March 30, 2023. https://edsource.org/2023/social-media-has-some-surprising-upsides-for-teenage-girls-study-finds/687726.

8. Thomas Wilde, "Why Avatar-Based Social Gaming App Rec Room Doesn't Consider Itself a 'Metaverse' Company," *Geek-Wire* (December 29, 2022). https://www.geekwire.com/2022/why-avatar-based-social-gaming-app-rec-room-doesnt-consider-itself-a-metaverse-company/.

9. Megan Leonhardt, "Meet a Millennial Who Is Turning 40, Starting Yet Another New Career and Has $47,000 in Debt. 'I've Worked Very Hard and It Didn't Pay off. It Feels Very Unfair,'" *Fortune* (February 20, 2022). https://fortune.com/2022/02/20/millennial-turning-40-starting-new-career-carrying-debt/#:~:text=Overall%2C%20the%20average%20millennial%20carries,debt%20averages%20%24255%2C527%20per%20person.

10. "'Entitled' Gen Z Is the Most Difficult Generation in the Workplace: Poll," *New York Post* on Twitter, April 21, 2023. https://twitter.com/nypost/status/1649538595877691392.

11. Lisa Costa and Michael Torres, "Case Study: How and Why the United States Space Force Is Building a Digital Twin," presentation at The Economist Metaverse Summit, San Jose, California, October 26, 2022.

12. BrainyQuote, "Pearl S. Buck Quotes," n.d. https://www.brainyquote.com/quotes/pearl_s_buck_161681.

Chapter 2

1. Michele Debczak, "Watch Keanu Reeves Describe 'The Matrix,'" *Mental Floss* (December 17, 2021). https://www.mentalfloss.com/article/653620/keanu-reeves-describes-matrix-to-unimpressed-teens.

2. "Razorfish Study Finds 52% of Gen Z Gamers Feel More Like Themselves in the Metaverse Than in Real Life," *Business Wire* (April 19, 2022). https://www.businesswire.com/news/home/20220419005232/en/.

3. Mike Boland, "How Do Consumers Feel About the Metaverse?" AR Insider, September 13, 2021. https://arinsider.co/2021/09/13/how-do-consumers-feel-about-the-metaverse/.

4. Everyrealm, "Immersive Media Market Overview," *Everyrealm* (blog), February 21, 2023. https://everyrealm.com/blog/reports/immersive-media-market-overview?utm_source=Everyworld&utm_campaign=1eaee35f35-EMAIL_CAMPAIGN_2023_02_09_01_22_COPY_01&utm_medium=email&utm_term=0_-65af8b81c5-%5BLIST_EMAIL_ID%5D.

5. Saeed Saeed, "Who Is Issam Alnajjar? The Jordanian Teen Singer Proving More Popular on Spotify Than Justin Bieber," *The National* (February 3, 2021). https://www.thenationalnews.com/arts-culture/music/who-is-issam-alnajjar-the-jordanian-teen-singer-proving-more-popular-on-spotify-than-justin-bieber-1.1158308.

6. Yarimar Bonilla, "Opinion | Bad Bunny Is [Winning in Non-English]," *New York Times* (February 11, 2023). https://www.nytimes.com/2023/02/11/opinion/bad-bunny-non-english-grammys.html?ref=refind.

7. Ritz Campbell, "MrBeast, Now in Spanish! Say Hello to Multi-Language Audio on YouTube," *YouTube Official Blog*, February 23, 2023. https://blog.youtube/news-and-events/multi-language-audio-mrbeast-interview/?tpcc=nldatasheet.

8. David Rose, *SuperSight* (Dallas: BenBella Books, 2021), 136.

9. Marius Rubo, Nadine Messerli, and Simone Munsch, "The Human Source Memory System Struggles to Distinguish Virtual Reality and Reality," *Computers in Human Behavior Reports* 4 (August 1, 2021): 100111. https://doi.org/10.1016/j.chbr.2021.100111.

Chapter 3

1. Mary Bellis, "Do You Know Who Invented the Selfie?" ThoughtCo, January 3, 2020. https://www.thoughtco.com/who-invented-the-selfie-1992418.

2. Allure, "The Evolution of the Selfie," *Allure* (February 23, 2021). https://www.allure.com/story/social-networks-impact-on-beauty.

3. Mariko Oi, "Is This the World's Largest Virtual Fashion Show?" *BBC News* (December 23, 2021). https://www.bbc.com/news/business-59558921.

4. Conversation with Shawn Whiting, March 14, 2023.

5. Remy Tumin, "Same-Sex Couple Households in U.S. Surpass One Million," *New York Times* (December 2, 2022). https://www.nytimes.com/2022/12/02/us/same-sex-households-census.html.

6. Kim Parker, Nikki Graf, Ruth Igielnik, and Travis Mitchell, "Generation Z Looks a Lot Like Millennials on Key Social and Political Issues," Pew Research Center's Social & Demographic Trends Project, September 17, 2019. https://www.pewresearch.org/social-trends/2019/01/17/generation-z-looks-a-lot-like-millennials-on-key-social-and-political-issues/.

7. Koh Mochizuki, "Indiana Fifth Graders Stage Walkout to Protest Anti-LGBTQ Bill – and the Internet Is Cheering," Second Nexus, March 25, 2023. https://secondnexus.com/indiana-fifth-graders-stage-walkout.

8. Leila Brillson, "Lily-Rose Depp & Harley Quinn Smith Are BFFs On and Off Screen," *Nylon* (February 3, 2016). https://www.nylon.com/articles/lily-rose-depp-harley-quinn-smith-interview.

9. "Gen Z Goes Beyond Gender Binaries in New Innovation Group Data," Wunderman Thompson, March 11, 2016. https://www.wundermanthompson.com/insight/gen-z-goes-beyond-gender-binaries-in-new-innovation-group-data.

10. Mark Jacobs, "Jaden Smith: Fresh Prince of the Future | Cover Shoot and Interview for GQ Style S/S '16," *British GQ* (March 11, 2016). https://www.gq-magazine.co.uk/article/jaden-smith-gq-style-cover-preview.

11. Leila Brillson, "Lily-Rose Depp & Harley Quinn Smith Are BFFs On and Off Screen."

12. Kurt Wagner, "Meta to Stop Letting Advertisers Target Teens by Gender," Bloomberg.com, January 10, 2023. https://www.bloomberg.com/news/articles/2023-01-10/meta-to-stop-letting-advertisers-target-teens-by-gender?leadSource=uverify%20wall.

13. Lucy Maguire, "Has the Gen Z Gender-Neutral Store Finally Arrived?" Vogue Business, October 9, 2020. https://www.voguebusiness.com/companies/has-the-gen-z-gender-neutral-store-finally-arrived.

14. Lucy Maguire, "Has the Gen Z Gender-Neutral Store Finally Arrived?"

15. Nazanin Lankarani, "Taking Gender Out of Watches," *New York Times* (January 27, 2022). https://www.nytimes.com/2022/01/27/fashion/watches-gender-zenith.html.

16. Lauren Thomas, "Lines Between Men's and Women's Fashion Are Blurring as More Retailers Embrace Gender-Fluid Style," CNBC (June 9, 2021). https://www.cnbc.com/2021/06/09/gender-fluid-fashion-booms-retailers-take-cues-from-streetwear-gen-z.html.

17. David Eagleman, *Livewired: The Inside Story of the Ever-Changing Brain* (Edinburgh: Canongate Books, 2020), 134.
18. Nick Yee and Jeremy N. Bailenson, "The Proteus Effect: The Effect of Transformed Self-Representation on Behavior," *Human Communication Research* 33, no. 3 (July 1, 2007): 271–290. https://doi.org/10.1111/j.1468-2958.2007.00299.x.
19. David Eagleman, *Livewired*, 135.
20. Sherry Turkle, *Alone Together: Why We Expect More from Technology and Less from Each Other* (New York: Basic Books, 2011), 272.
21. David Eagleman, *Livewired*, 136.
22. Dalila Burin, Gabriele Cavanna, Daniela Rabellino, Yuka Kotozaki, and Ryuta Kawashima, "Neuroendocrine Response and State Anxiety Due to Psychosocial Stress Decrease After a Training with Subject's Own (but Not Another) Virtual Body: An RCT Study," *International Journal of Environmental Research and Public Health* 19, no. 10 (May 23, 2022): 6340. https://doi.org/10.3390/ijerph19106340.
23. Jih-Hsuan Tammy Lin and Dai-Yun Wu, "Exercising with Embodied Young Avatars: How Young vs. Older Avatars in Virtual Reality Affect Perceived Exertion and Physical Activity Among Male and Female Elderly Individuals," *Frontiers in Psychology* 12 (October 25, 2021). https://doi.org/10.3389/fpsyg.2021.693545.
24. Sandra Blakeslee, "Clothes and Self-Perception," *New York Times* (April 2, 2012). https://www.nytimes.com/2012/04/03/science/clothes-and-self-perception.html.
25. Sandra Blakeslee, "Clothes and Self-Perception."
26. Hope, Jessica. "Timeline: The Rise and Fall of the Roman Games," Historyrevealed.com, October 3, 2018. https://www.historyrevealed.com/eras/ancient-rome/timeline-the-rise-and-fall-of-the-roman-games/.
27. Nir Eyal, *Hooked: How to Build Habit-Forming Products* (Penguin, 2014).

Chapter 4

1. Werner Geyser, "Top Female Twitch Streamers – The Rise of Female Streamers in Gaming," Influencer Marketing Hub, May 24, 2022. https://influencermarketinghub.com/top-female-twitch-streamers/.

2. Emma Sandler, "MAC Cosmetics Targets Gamers with Twitch-Con Sponsorship," *Glossy* (September 26, 2019). https://www.glossy.co/beauty/mac-cosmetics-targets-gamers-with-twitchcon-sponsorship/#:~:text=When%20attendees%20to%20the%20annual%20video%20game%20convention,like%20energy%20drinks%20and%20indie%20video%20game%20studios.

3. J.C.R. Licklider and Robert W. Taylor, "The Computer as a Communication Device," *Science and Technology* (April 1968). https://internetat50.com/references/Licklider_Taylor_The-Computer-As-A-Communications-Device.pdf.

4. Reuben J. Thomas, "Online Dating, Now the Most Common Way for Couples to Meet, Is Desegregating America," *NBC News* (October 6, 2019). https://www.nbcnews.com/think/opinion/online-dating-now-most-common-way-couples-meet-desegregating-america-ncna1062731.

5. "Meet the Drag Queen Revolutionizing the Streaming World on Twitch," *NBC News* (September 29, 2019). https://www.nbcnews.com/feature/nbc-out/meet-drag-queen-revolutionizing-streaming-world-twitch-n1060021.

6. "Meet the Drag Queen Revolutionizing the Streaming World on Twitch."

7. "Meet the Drag Queen Revolutionizing the Streaming World on Twitch."

8. S. K., "Virtual Influencer Lil Miquela's New Breakup Single 'Speak Up': A Breakdown," *Virtual Humans*, March 16, 2020. https://www.virtualhumans.org/article/virtual-influencer-lil-miquelas-new-breakup-single-speak-up-a-breakdown.

9. Jael Goldfine, "Lil Miquela Has 'Consciously Uncoupled' from Her Human Boyfriend," *PAPER* (March 6, 2020). https://www.papermag.com/lil-miquela-break-up-2645414969.html#rebelltitem5.

10. "Who Is Miquela Sousa? @lilmiquela, Explained," *Virtual Humans*, n.d. https://www.virtualhumans.org/human/miquela-sousa.

11. Kaitlyn Tiffany, "Lil Miquela and the Virtual Influencer Hype, Explained," *Vox* (June 3, 2019). https://www.vox.com/the-goods/2019/6/3/18647626/instagram-virtual-influencers-lil-miquela-ai-startups.

12. "The 25 Most Influential People on the Internet," *Time* (June 30, 2018). https://time.com/5324130/most-influential-internet/.

13. Maria Lewczyk, "The Appeal of Online Personas: Why People Love Virtual Influencers," *Virtual Humans*, June 30, 2021. https://www.virtualhumans.org/article/online-personas-and-why-people-love-virtual-influencers.

14. Charles Trepany, "The Robot Invasion Has Begun: Meet the Computer-Generated Influencers Taking Over Instagram," *USA Today* (February 3, 2021). https://www.usatoday.com/story/life/2019/10/16/cgi-influencers-blur-line-between-reality-and-fantasy-instagram-advertising/3790471002/.

15. Midori Sugama, "IKEA Shares a Glimpse of Home Life with Imma: Japan's First Virtual Model," Designboom, September 11, 2020. https://www.designboom.com/technology/ikea-home-life-with-imma-japans-first-virtual-model-09-11-2020/.

16. "CGI Influencers: Bot or Not?" TBH, November 2019. https://grin.co/wp-content/uploads/2021/10/Fullscreen_CGI-Influencers_Bot-Or-Not.pdf.

17. "Study: Many People Prefer to Speak About Negative Experiences with a Virtual Reality Avatar," News-Medical.Net, January 5, 2022. https://www.news-medical.net/news/20220105/Study-Many-people-prefer-to-speak-about-negative-experiences-with-a-virtual-reality-avatar.aspx.

18. Mixed Reactions, "Gamers Reaction to Arthur Telling Sister He's Afraid on *Red Dead Redemption 2*," YouTube, May 22, 2022. https://www.youtube.com/watch?v=fFO6QB51QSU.

19. Devin Ellis Friend, *"Red Dead Redemption 2* Is Too Sad to Replay,"* ScreenRant, May 4, 2022. https://screenrant.com/rdr2-story-sad-replay-arthur-morgan-dutch-marston/.

20. "Okay This Game Made Me Cry: Life Is Strange™ General Discussions," Steam Community Board, February 16, 2018. https://steamcommunity.com/app/319630/discussions/0/2860219962086308349/.

21. Max Dyckhoff, "Ellie: Buddy AI in *The Last of Us*," Presentation at the Game Developers Conference, March 18, 2014. https://www.gdcvault.com/play/1020364/Ellie-Buddy-AI-in-The.

22. Victor Tangermann, "ChatGPT Mod Allows NPCs in Skyrim to Make Up New Dialogue," *Futurism* (May 1, 2023). https://futurism.com/the-byte/chatgpt-mod-npcs-skyrim.

23. Sangeeta Singh-Kurtz, "The Man of Your Dreams," The Cut, March 10, 2023. https://www.thecut.com/article/ai-artificial-intelligence-chatbot-replika-boyfriend.html.

24. Amelia Hill, "'We're on Permanent Catch-up': How Covid Has Changed Young Britons' Lives," *Guardian* (January 29, 2023). https://www.theguardian.com/society/2023/jan/29/were-on-permanent-catch-up-how-covid-has-changed-young-britons-lives.

Chapter 5

1. Amelia Hill, "'We're on Permanent Catch-up': How Covid Has Changed Young Britons' Lives," *Guardian* (January 29, 2023). https://www.theguardian.com/society/2023/jan/29/were-on-permanent-catch-up-how-covid-has-changed-young-britons-lives.

2. Lili Loofbourow, "Volodymyr Zelensky Is Performing an Entirely Different Type of Politics," *Slate* (December 27, 2022). https://slate.com/news-and-politics/2022/12/volodymyr-zelenksy-performance-of-politics-effective.html.

3. Kieran Press-Reynolds, "Zelensky Fans Flood Social Media with Fancams and Thirsty Posts, Creating a Controversial Online Obsession," *Insider* (March 2, 2022). https://www

.insider.com/volodymyr-zelensky-ukrainian-president-selfie-videos-fancams-tiktok-twitter-2022-3#:~:text=Ukrainian%20President%20Volodymyr%20Zelensky%20has%20become%20an%20online,created%20fancams%20and%20memes%2C%20prompting%20disagreements%20about%20tone.

4. Paige Leskin, "American Kids Want to Be Famous on YouTube, and Kids in China Want to Go to Space: Survey," *Business Insider* (July 17, 2019). https://www.businessinsider.com/american-kids-youtube-star-astronauts-survey-2019-7.

5. Alessandra Conti, "Six Celebrity Secrets to Being Charismatic," *Forbes* (July 10, 2020). https://www.forbes.com/sites/forbesbusinesscouncil/2020/07/10/six-celebrity-secrets-to-being-charismatic/?sh=19306f651494.

6. "2018 Plastic Surgery Statistics Report," American Society of Plastic Surgeons, n.d, 7. https://www.plasticsurgery.org/documents/News/Statistics/2018/plastic-surgery-statistics-full-report-2018.pdf.

7. Casey Newton, "Why BeReal Is Breaking Out," The Verge, July 20, 2022. https://www.theverge.com/2022/7/20/23271184/bereal-app-popularity-memes-tiktok-time-to.

8. Pesala Bandara, "Instagram Launches Its CopyCat BeReal Feature 'Candid Stories,'" PetaPixel, December 13, 2022. https://petapixel.com/2022/12/13/instagram-launches-its-copycat-bereal-feature-candid-stories/.

9. Kait Hanson, "New App Gas Creates Positive Social Network for Teens," *Today* (January 10, 2023). https://www.today.com/parents/family/what-is-gas-app-rcna64969.

10. Geeks of Color, "Rihanna's Interpreter Is the Real #Superbowl MVP," Twitter, February 12, 2023. https://twitter.com/GeeksOfColor/status/1624953250607951874

11. Ryan Aguirre, "I Really Wish This ASL Interpreter Got Some National Airtime," Twitter, February 12, 2023. https://twitter.com/aguirreryan/status/1624960501242540032?s=20

12. Rohan Mattu, "Deaf Performer and Bowie State Student Justina Miles Lands *British Vogue* Cover," *CBS News* (April 21,

2023). https://www.cbsnews.com/baltimore/news/deaf-perfor mer-and-bowie-state-student-justina-miles-lands-british-vogue-cover/.

13. Gianna Prudente, "This College Student Shares the Power of Building a Personal Brand," LinkedIn, February 22, 2023. https://www.linkedin.com/pulse/college-student-shares-power-building-personal-brand-gianna-prudente/?midToken= AQGmi0B9i_Ca2A&midSig=0dLdnD8X3pwaE1&trk=eml -email_series_follow_newsletter_01-newsletter_hero_banner-0-open_on_linkedin_cta&trkEmail=eml-email_ series_follow_newsletter_01-newsletter_hero_banner-0-open_ on_linkedin_cta-null-10dm5w~leftq1b1~s2-null-null&eid=10dm5w-leftq1b1-s2.

14. "How the Young Spend Their Money," *The Economist* (January 16, 2023). https://www.economist.com/business/2023/01/16/ how-the-young-spend-their-money.

15. "GenZ's Brutal Feedback on Your Marketing," Knit, June 2022. https://goknit.com/wp-content/uploads/2022/06/Download-Knit-Gen-Zs-Brutal-Feedback-on-Your-Marketing-Report-June-20-2022.pdf

Chapter 6

1. Deep Patel, "How Gen Z Will Shape the Future of Business," *Forbes* (April 18, 2017). https://www.forbes.com/sites/deeppa tel/2017/04/18/how-gen-z-will-shape-the-future-of-business/?sh=461a314d76e8.

2. Casey Baseel, "Nearly 70 Percent of Young Japanese Women Self-Identify as Otaku in Survey," *SoraNews24 – Japan News* (July 17, 2018). https://soranews24.com/2018/07/17/nearly-70-percent-of-young-japanese-women-self-identify-as-otaku-in-survey/.

3. Jake Sturmer, Yumi Asada, and Deborah Richards, "In Japan, a Million People Have Shut Themselves in Their Rooms. One Mother Is Helping Them Come Out," *ABC News Australia* (February 16, 2022). https://www.abc.net.au/news/2022-02-17/ hikikomori-seiko-goto-japan/100792330.

4. "Japan's 'Hikikomori' Population Could Top 10 Million," Nippon.com, September 17, 2019. https://www.nippon.com/en/japan-topics/c05008/japan%E2%80%99s-hikikomori-population-could-top-10-million.html.
5. Takeshi Kamiya, "Japan, S. Korea Jointly Tackle Problem of Social 'Shut-Ins,'" *Asahi Shimbun* (September 9, 2019). https://www.asahi.com/ajw/articles/13058624.
6. Neil Postman, "Five Things We Need to Know About Technological Change" (March 28, 1998). https://web.cs.ucdavis.edu/~rogaway/classes/188/materials/postman.pdf.
7. Neil Postman, *Amusing Ourselves to Death: Public Discourse in the Age of Show Business* (London: Penguin, 2005). First published 1985.
8. John Hanke, "The Metaverse Is a Dystopian Nightmare. Let's Build a Better Reality," Niantic, August 10, 2021. https://nianticlabs.com/news/real-world-metaverse?hl=en.

Chapter 7

1. Nick Pinkerton, "The Golden Age of NYC Public-Access TV at MOMI," *Village Voice* (May 11, 2017). https://www.villagevoice.com/2011/02/09/the-golden-age-of-nyc-public-access-tv-at-momi/.
2. Joel Topcik, "The Ugly George Hour of Truth, Sex & Convergence," *Broadcasting & Cable* (blog), January 10, 2009. https://www.nexttv.com/blog/ugly-george-hour-truth-sex-convergence-119966.
3. "How the Young Spend Their Money," *The Economist* (January 16, 2023). https://www.economist.com/business/2023/01/16/how-the-young-spend-their-money.
4. Madhav, "Twitch Business Model | How Does Twitch Make Money?" *SEOAves* (blog), January 1, 2022. https://seoaves.com/twitch-business-model-how-does-twitch-make-money/.
5. "Twitch Affiliate Partner Program – Reviews, News and Ratings," Business of Apps, November 27, 2022. https://www.businessofapps.com/affiliate/twitch/#:~:text=That%20

depends%20on%20the%20streamer,about%20%24250%20 every%20100%20subscribers.

6. Jordan Rose, "Kid Cudi Co-Founded Interactive Live Performance App Encore Launches With $9 Million in Funding Raised," Complex, February 16, 2022. https://www.complex .com/music/kid-cudi-co-founded-interactive-live-perfor mance-app-encore-launches-9-million-funding.

7. NPR, "So You Want to Be an Influencer?" *NPR Up First* podcast (May 7, 2023). https://www.npr.org/transcripts/11743 45890.

8. NPR, "So You Want to Be an Influencer?"

9. "Fortnite World Cup Winner: 'Bugha' (Kyle Giersdorf) – Age 16 – Wins $3 Million at Fortnite World Cup," *CBS News* (July 29, 2019). https://www.cbsnews.com/news/fortnite-world-cup-winner-bugha-kyle-giersdorf-age-16-wins-3-million-at-fortnite-world-cup/.

10. Brian Feldman, "Before 'Old Town Road,' Lil Nas X Was a Tweetdecker," *Intelligencer* (April 5, 2019). https://nymag.com/ intelligencer/2019/04/lil-nas-x-was-a-popular-twitter-user-before-old-town-road.html.

11. @LilNasX, "country music is evolving," Twitter, December 2, 2018. https://twitter.com/LilNasX/status/106942490137 3730816?s=20.

12. Julius Young, "John Rich Speaks Out on Lil Nas X's 'Old Town Road' After Billy Ray Cyrus Hops on Remix: 'Let the Fans Decide,'" *Fox News* (April 5, 2019). https://www.foxnews.com/ entertainment/john-rich-speaks-out-on-lil-nas-xs-old-town-road-after-billy-ray-cyrus-hops-remix-let-the-fans-decide.

13. @LilNasX, "twitter please help me get billy ray cyrus on this," Twitter, December 4, 2018. https://twitter.com/LilNasX/status/ 1070125203345342464?s=20.

14. "Say Hello to the Snapchat Generation," Snapchat, 2022, 8. https://downloads.ctfassets.net/inb32lme5009/30vuj5q8bEp9T 136adWN86/0ba92cb714da2eeba3be032051b33dd7/The_ Snapchat_Generation_North_America.pdf.

15. "Into Z Future: Understanding Gen Z, The Next Generation of Super Creatives," J. Walter Thompson Intelligence, 2019, 44. https://assets.ctfassets.net/inb32lme5009/5DFlqKVGIdmAu7X6btfGQt/44fdca09d7b630ee28f5951d54feed71/Into_Z_Future_Understanding_Gen_Z_The_Next_Generation_of_Super_Creatives_.pdf.

16. "Insights from Our '2022 Metaverse Fashion Trends' Report," *Roblox* (blog), November 1, 2022. https://blog.roblox.com/2022/11/insights-from-our-2022-metaverse-fashion-trends-report/.

17. "Insights from Our '2022 Metaverse Fashion Trends' Report," *Roblox* (blog).

18. @UjuAnya, "A laptop arrived at our house," Twitter, May 12, 2023. https://twitter.com/ujuanya/status/1657040823979061248?s=48&t=7vuI8kxA8e8hFdZb6wno_g.

19. @UjuAnya, "Bruh. I grilled him like the CIA," Twitter, May 12, 2023. https://twitter.com/UjuAnya/status/1657042354077945857.

20. @DivYank,"Even being 25 in this world," Twitter, May 12, 2023. https://twitter.com/divyank_diwakar/status/1657222010836258816.

21. Katharine K. Zarrella, "Pajamas Are the New Sweatpants – How to Wear Them Everywhere," *Wall Street Journal* (January 16, 2021). https://www.wsj.com/articles/pajamas-are-the-new-sweatpantshow-to-wear-them-everywhere-11610773201.

22. Carla Mozée, "A 28-Year-Old on Asia's Top Metaverse Platform Makes Six Figures as a Virtual Fashion Influencer," *Markets Insider* (December 25, 2021). https://markets.businessinsider.com/news/currencies/metaverse-fashion-designer-influencer-monica-quin-zepeto-six-figure-income-2021-12.

23. Darius McQuaid, "CEO of Fortnite Developer Said He Will Not 'Touch NFTs,'" Currency.com, September 28, 2021. https://currency.com/ceo-of-fortnite-developer-said-he-will-not-touch-nfts#:~:text=The%20CEO%20of%20Epic%20Games%2C%20the%20American%20video,tangled%20up%20with%20an%20intractable%20mix%20of%20scams%E2%80%9D.

24. Lora Jones and Laura Heighton-Ginns, "Roblox: 'We Paid off Our Parents' Mortgage Making Video Games,'" *BBC News* (March 12, 2021). https://www.bbc.com/news/business-56354253.

25. "Roblox Corporation Developer and Creator Breakdown 2022, by Rewards," Statista, April 3, 2023. https://www.statista.com/statistics/1191213/roblox-developer-creator-breakdown-rewards/.

26. Andy Chalk, "Roblox Faces New Allegations of Being Unsafe for Children," *PC Gamer* (December 14, 2021). https://www.pcgamer.com/roblox-faces-new-allegations-of-being-unsafe-for-children/.

27. Brian Dean, "Roblox User and Growth Stats 2022," Backlinko, October 10, 2022. https://backlinko.com/roblox-users.

28. Simon Parkin, "The Trouble with Roblox, the Video Game Empire Built on Child Labour," *Guardian* (January 20, 2022). https://www.theguardian.com/games/2022/jan/09/the-trouble-with-roblox-the-video-game-empire-built-on-child-labour.

29. Quintin Smith, "Investigation: How Roblox Is Exploiting Young Game Developers," YouTube, August 19, 2021. https://www.youtube.com/watch?v=_gXlauRB1EQ, and "Roblox Pressured Us to Delete Our Video. So We Dug Deeper," YouTube, December 13, 2021. https://www.youtube.com/watch?v=vTMF6xEiAaY.

30. "How Roblox Exploits Children, with Quintin Smith," *Tech Won't Save Us* podcast (January 6, 2022). https://podcasts.apple.com/us/podcast/how-roblox-exploits-children-w-quintin-smith/id1507621076?i=1000547080283.

Chapter 8

1. Amelia Hill, "'We're on Permanent Catch-up': How Covid Has Changed Young Britons' Lives," *Guardian* (January 29, 2023). https://www.theguardian.com/society/2023/jan/29/were-on-permanent-catch-up-how-covid-has-changed-young-britons-lives.

2. "Failure Drives Innovation, According to EY Survey on Gen Z," Cision PR Newswire, September 18, 2018. https://www.prnewswire.com/news-releases/failure-drives-innovation-according-to-ey-survey-on-gen-z-300714436.html.

3. Gianna Prudente, "Bridging the Gap: How to Navigate a Multigenerational Workplace Early on in Your Career," LinkedIn, January 18, 2023. https://www.linkedin.com/pulse/bridging-gap-how-navigate-multigenerational-workplace-gianna-prudente/?midToken=AQGmi0B9i_Ca2A&midSig=2VWKozH-LYKWA1&trk=eml-email_series_follow_newsletter_01-newsletter_hero_banner-0-open_on_linkedin_cta&trkEmail=eml-email_series_follow_newsletter_01-newsletter_hero_banner-0-open_on_linkedin_cta-null-10dm5w~ld1tf081~49-null-null&eid=10dm5w-ld1tf081-49.

4. Fabiana Buontempo, "Things Parents Don't Understand in 2023," *BuzzFeed* (January 28, 2023). https://www.buzzfeed.com/fabianabuontempo/young-adult-things-parents-dont-understand.

5. Louis Uchitelle, "Job Insecurity of Workers Is a Big Factor in Fed Policy," *New York Times* (February 27, 1997). https://www.nytimes.com/1997/02/27/business/job-insecurity-of-workers-is-a-big-factor-in-fed-policy.html.

6. Guy Standing, "Meet the Precariat, the New Global Class Fuelling the Rise of Populism," World Economic Forum, November 9, 2016. https://www.weforum.org/agenda/2016/11/precariat-global-class-rise-of-populism/.

7. Guy Standing, "Meet the Precariat, the New Global Class Fuelling the Rise of Populism."

8. David Marchese, "Thomas Piketty Thinks America Is Primed for Wealth Redistribution," *New York Times* (April 1, 2022). https://www.nytimes.com/interactive/2022/04/03/magazine/thomas-piketty-interview.html.

9. Ana Hernandez Kent and Lowell R. Ricketts, "Has Wealth Inequality in America Changed over Time? Here Are Key Statistics," *Federal Reserve Bank of St. Louis* (blog), December 2,

2020. Accessed May 4, 2023. https://www.stlouisfed.org/open-vault/2020/december/has-wealth-inequality-changed-over-time-key-statistics.

10. John Cassidy, "Piketty's Inequality Story in Six Charts," *New Yorker* (March 26, 2014). https://www.newyorker.com/news/john-cassidy/pikettys-inequality-story-in-six-charts.

11. David Marchese, "Thomas Piketty Thinks America Is Primed for Wealth Redistribution."

12. Jessica Dickler, "More Education Doesn't Always Get You More Money, Report Finds," *CNBC* (October 13, 2021). https://www.cnbc.com/2021/10/13/more-education-doesnt-always-get-you-more-money-report-finds.html.

13. Melanie Hanson, "Average Cost of College [2023]: Yearly Tuition + Expenses," Education Data Initiative, April 3, 2023. https://educationdata.org/average-cost-of-college.

14. Michael Mitchell, Michael Leachman, and Kathleen Masterson, "A Lost Decade in Higher Education Funding," Center on Budget and Policy Priorities, August 23, 2017. https://www.cbpp.org/research/state-budget-and-tax/a-lost-decade-in-higher-education-funding.

15. "2022 Instagram Trend Report," Instagram, December 13, 2021. https://about.instagram.com/blog/announcements/instagram-trends-2022.

16. Willy Staley, "How Many Billionaires Are There, Anyway?" *New York Times* (April 7, 2022). https://www.nytimes.com/2022/04/07/magazine/billionaires.html?action=click&pgtype=Interactive&state=default&module=styln-money-issue&variant=show®ion=BELOW_MAIN_CONTENT&block=storyline_flex_guide_recirc.

17. Mindy Shoss, Shiyang Su, Ann Schlotzhauer, and Nicole Carusone, "Job Insecurity Harms Both Employees and Employers," *Harvard Business Review* (September 26, 2022). https://hbr.org/2022/09/job-insecurity-harms-both-employees-and-employers.

18. Mindy Shoss et al., "Job Insecurity Harms Both Employees and Employers."

19. Dana George, "5 Reasons Dave Ramsey Says You Should 'Act Your Wage,'" The Motley Fool, October 30, 2022. https://www.fool.com/the-ascent/personal-finance/articles/5-reasons-dave-ramsey-says-you-should-act-your-wage/#:~:text=According%20to%20the%20ever-entertaining%20%28and%20often%20crude%29%20Urban,acting%20your%20wage%20means%20asserting%20boundaries%20at%20work.

20. DeAndre Brown,"The First Day Back from the Holiday Should Only Be Used for Checking Emails!" TikTok, n.d. https://www.tiktok.com/@imdrebrown/video/7184946665609235758.

21. Fabiana Buontempo, "Things Parents Don't Understand in 2023," BuzzFeed (January 28, 2023). https://www.buzzfeed.com/fabianabuontempo/young-adult-things-parents-dont-understand.

22. Jasmine McCall, "31-year-old makes $105,000 a month in passive income from her side hustle: 'I work just 2 hours a day,'" CNBC on MSN.com (April 19, 2023). https://www.msn.com/en-us/money/careers/31-year-old-makes-105-000-a-month-in-passive-income-from-her-side-hustle-i-work-just-2-hours-a-day/ar-AA1a3V76?ocid=winp1taskbar&cvid=791c10b1427745aca9b9c11eb14eab95&ei=12.

23. Amelia Hill, "'We're on Permanent Catch-up.'"

24. Gianna Prudente, "This Gen Zer Is Sharing How Being a Fashion Content Creator Has Benefited Her Finance Career," LinkedIn, March 8, 2023. https://www.linkedin.com/pulse/gen-zer-sharing-how-being-fashion-content-creator-has-gianna-prudente/?trackingId=UYueuRtNS3qBTTDpeyfFqA%3D%3D.

25. Jennifer Ortakales Dawkins, "I'm a College Student Selling $13K Annually on Poshmark. Here's the Daily Routine I Follow to Juggle Resale and School," Business Insider (November 21, 2022). https://www.businessinsider.com/how-to-sell-on-poshmark-make-money-gen-z-entrepreneur-2022-7#i-woke-up-at-5-am-checked-emails-and-did-my-homework-1.

26. George Anders, "Is Gen Z the Boldest Generation? Its Job-Hunt Priorities Are off the Charts," LinkedIn, February 9, 2022.https://www.linkedin.com/pulse/gen-z-boldest-generation-

its-job-hunt-priorities-off-charts-anders/?trackingId=pwWrCQ
Q1SiG9Yds3hH8gUg%3D%3D.

27. Brooklyn White, "On Average, Gen Z Is Staying at One Job for Just over 2 Years. This Is Why," Girls United, December 17, 2021.https://girlsunited.essence.com/article/gen-z-job-hopping-feature/#:~:text=Gen%20Z%2C%20which%20caps%20off%20around%2023-24%20years,next%20gig%2C%20one%202021%20study%20from%20CareerBuilder%20says.

28. Brooklyn White, "On Average, Gen Z Is Staying at One Job for Just over 2 Years."

29. "The State of Internal Mobility and Employee Retention Report," Lever, March 22, 2023. https://www.lever.co/research/2022-internal-mobility-and-employee-retention-report/?utm_medium=pr&utm_source=news&utm_campaign=leverbrand&utm_content=job%20crafting%20reporttps://www.businessinsider.com/labor-shortage-slow-up-great-reshuffle-gen-z-demands-workplace-2021-11.

30. Vaishali Sabhahit, "Adobe's Future Workforce Study Reveals What the Next Generation Workforce Is Looking for in the Workplace," *Adobe Blog*, January 24, 2023. https://blog.adobe.com/en/publish/2023/01/24/adobes-future-work force-study-reveals-what-next-generation-workforce-looking-for-in-workplace.

31. Alicia Adamczyk, "A 26-Year-Old Quit His Job in Advertising Because He Can Make More as a TikTok Creator – Here's How He Did It," *Fortune* (February 19, 2023). https://fortune.com/2023/02/19/how-to-quit-job-become-full-time-tik-tok-creator/?utm_source=email&utm_medium=newsletter&utm_campaign=data-sheet&utm_content=2023022119pm&tpcc=nldatasheet.

32. "2022 Retention Report: How Employers Caused the Great Resignation," Work Institute, 2022. https://info.workinstitute.com/2022-retention-report.

33. George Anders, "Is Gen Z the Boldest Generation?"

34. Angela Littlefield, "'They Kept Muting Me to Talk to Each Other': Job Hunter Rescinds Application for 6-Figure Salary Job after Spotting Red Flags during Hiring Process," The Daily

Dot, April 20, 2023. https://www.dailydot.com/news/job-hunter-rescinds-application-red-flags/.

35. "What Gen-Z Graduates Want from Their Employers," *The Economist* (July 21, 2022). https://www.economist.com/business/2022/07/21/what-gen-z-graduates-want-from-their-employers?utm_content=ed-picks-article-link-3&etear=nl_special_3&utm_campaign=a.coronavirus-special-edition&utm_medium=email.internal-newsletter.np&utm_source=salesforce-marketing-cloud&utm_term=1/21/2023&utm_id=1456534.

36. Bridget Scanlon and Neal Sivadas, "Does Generation Z Want to Work from Home?" JUV Consulting, October 9, 2020. https://www.juvconsulting.com/add-date-does-generation-z-want-to-work-from-home/#:~:text=When%20asked%20whether%20Gen%20Zers%20would%20like%20to,work%20atmosphere%20of%20all%20worklife%2C%20and%20no%20balance.

37. "What Gen-Z Graduates Want from Their Employers."

38. Derek Thompson, "Quit Your Job," *The Atlantic* (November 5, 2014). https://www.theatlantic.com/business/archive/2014/11/quit-your-job/382402/.

39. Jon Younger, "Freelance Marketplaces, Are You Prepared to Be Disrupted?" *Forbes* (September 11, 2022). https://www.forbes.com/sites/jonyounger/2022/09/11/freelance-marketplaces-are-you-prepared-to-be-disrupted/?sh=1b4e2fd55a96.

40. "State of Independence in America Report 2022," MBO Partners, April 4, 2023. https://www.mbopartners.com/state-of-independence/.

41. "Riga Innovatively Engages Young People in Urban Planning and City Exploration," Riga City Council, October 19, 2022. https://www.riga.lv/en/article/riga-innovatively-engages-young-people-urban-planning-and-city-exploration.

42. Ali Shutler, "London Mayor Sadiq Khan Joins 'Minecraft' and Asks Players to Redesign Croydon," NME, January 26, 2023. https://www.nme.com/news/gaming-news/london-mayor-sadiq-khan-joins-minecraft-and-asks-players-to-redesign-croydon-3387429.

43. Sergio Goschenko, "Recruiting Agencies in Japan Are Turning to the Metaverse," *Bitcoin News* (January 30, 2023). https://news.bitcoin.com/recruiting-agencies-in-japan-are-turning-to-the-metaverse/?mkt_tok=NjczLVBISy05NDgAAAGJseJH6hS skZPZddRmv1kOcEqxjxX10klMkatpqplX36r5h3ILy 2WzMT5Bu_VK_SW4SucUCDljyOyH3tRW2e6FyxhRVDZ Kz6-WM646ZqQ.

44. Renee Dudley and Daniel Golden, *The Ransomware Hunting Team: A Band of Misfits' Improbable Crusade to Save the World from Cybercrime* (New York: Farrar, Straus and Giroux, 2022), 195.

45. Colin Campbell, "Gen Z Is Ready to Tackle Gaming's Biggest Challenges," GameDaily.biz, March 13, 2023. https://www.gamedaily.biz/gen-z-is-ready-to-tackle-gamings-biggest-challenges/.

46. Carol Pogash, "On TikTok and YouTube, Here's Why Quitting Videos Go Viral," *New York Times* (March 11, 2023). https://www.nytimes.com/2023/03/09/business/quitting-videos-viral-tiktok-youtube.html.

Chapter 9

1. Taylor Knight, "Keke Palmer Gives Birth – in 'Realistic' 'Sims' Game," *New York Post* (February 9, 2023). https://nypost.com/2023/02/09/keke-palmer-gives-birth-in-realistic-sims-game/.

2. Sarah Perez, "Google Exec Suggests Instagram and TikTok are Eating into Google's Core Products, Search and Maps," Tech-Crunch, July 12, 2022. https://techcrunch.com/2022/07/12/google-exec-suggests-instagram-and-tiktok-are-eating-into-googles-core-products-search-and-maps/?tpcc=nldatasheet.

Chapter 10

1. "A Tech Worker Ammaar Reshi Is Selling a Children's Book He Made Using AI. Professional Illustrators Are Pissed," *Bluesyemre* (blog), December 20, 2022. https://bluesyemre

.com/2022/12/20/a-tech-worker-ammaar-reshi-is-selling-a-childrens-book-he-made-using-ai-professional-illustrators-are-pissed/#:~:text=Ammaar%20Reshi%2C%2028%2C%20has%20been%20fascinated%20by%20technology,UK%2C%20where%20Reshi%20studied%20computer%20science%20in%20London.

2. Ammaar Reshi, "I spent the weekend playing with ChatGPT," Twitter, December 9, 2022. https://twitter.com/ammaar/status/1601284293363261441

3. Bluesyemre, "A Tech Worker Ammaar Reshi Is Selling a Children's Book."

4. Kalley Huang, "Alarmed by A.I. Chatbots, Universities Start Revamping How They Teach," *New York Times* (January 16, 2023). https://www.nytimes.com/2023/01/16/technology/chatgpt-artificial-intelligence-universities.html.

5. Pia Ceres, "ChatGPT Is Coming for Classrooms. Don't Panic," *WIRED* (January 26, 2023). https://www.wired.com/story/chatgpt-is-coming-for-classrooms-dont-panic/?redirectURL=https%3A%2F%2Fwww.wired.com%2Fstory%2Fchatgpt-is-coming-for-classrooms-dont-panic%2F.

6. Nicholas Carr, (2010). *The Shallows: What the Internet Is Doing to Our Brains* (New York: Norton), 117.

7. Nicholas Carr, *The Shallows*, 141.

8. "Behind the Scenes with Psychologist Patricia Greenfield," UCLA Newsroom, July 22, 2016. https://newsroom.ucla.edu/stories/behind-the-scenes-with-psychologist-patricia-greenfield.

9. Patricia M. Greenfield, 2014, *Mind and Media: The Effects of Television, Video Games, and Computers* (New York: Psychology Press), 97. First published 1984.

10. Elon Musk, "It's a New World," Twitter, January 4, 2023. https://twitter.com/elonmusk/status/1610849544945950722?s=20.

11. Pia Ceres, "ChatGPT Is Coming for Classrooms. Don't Panic," *WIRED* (January 26, 2023). https://www.wired.com/story/chatgpt-is-coming-for-classrooms-dont-panic/?redirectURL=https%3A%2F%2Fwww.wired.com%2Fstory%2Fchatgpt-is-coming-for-classrooms-dont-panic%2F.

12. Stefan Popenici and Sharon Kerr. "Exploring the Impact of Artificial Intelligence on Teaching and Learning in Higher Education," *Research and Practice in Technology Enhanced Learning* 12, no. 1 (December 1, 2017). https://doi.org/10.1186/s41039-017-0062-8.
13. Greg Rosalsky, "This 22-Year-Old Is Trying to Save Us from ChatGPT before It Changes Writing Forever," *NPR* (January 17, 2023). https://www.npr.org/sections/money/2023/01/17/1149206188/this-22-year-old-is-trying-to-save-us-from-chatgpt-before-it-changes-writing-for.
14. Mike Cassidy, "Centaur Chess Shows Power of Teaming Human and Machine," *HuffPost* (December 7, 2017). https://www.huffpost.com/entry/centaur-chess-shows-power_b_6383606.
15. Shelly Palmer, "There Is a Profound Difference Between Losing Your Job to a Machine . . ." LinkedIn, March 27, 2023. Accessed May 3, 2023. https://www.linkedin.com/posts/shellypalmer_there-is-a-profound-difference-between-losing-activity-7046096336227573760-UBC_/.

Chapter 11

1. W. H. Auden, "Paul Bunyan: An Operetta in Two Acts and a Prologue" (New York: Faber, 1941).
2. Sara Colmenares, "The Treasure of 'El Dorado': The Guatavita Lagoon," *Sula Travel Agency* (blog), January 9, 2020. https://www.sula.com.co/blog/guatavita-lagoon/.
3. Taylor Locke, "Crypto Is 'the Future of Finance': Why Gen Z Is Ditching Traditional Investments—but with Caution," *CNBC* (June 22, 2021). https://www.cnbc.com/2021/06/22/gen-z-investing-in-cryptocurrency-btc-eth-and-meme-stocks-amc-gme.html.

Chapter 12

1. Stef W. Kight, "Gen Z Is Eroding the Power of Misinformation," *Axios* (September 15, 2020). https://www.axios.com/2020/09/15/gen-z-is-eroding-the-power-of-misinformation.

2. Li Miao, Jinyoung Im, Kevin Kam Fung So, and Yan Cao, "Post-Pandemic and Post-Traumatic Tourism Behavior," *Annals of Tourism Research* 95 (July 1, 2022): 103410. https://doi.org/10.1016/j.annals.2022.103410.

3. Cecily Mauran, "What Is Revenge Travel and Why Is Everyone Talking About It?" *Mashable* (August 27, 2022). https://mashable.com/article/what-is-revenge-travel-explainer.

4. Monica Pitrelli, "'Air Rage' Is Complicating Travel in North America and Europe – but Not so Much in Asia," *CNBC* (February 23, 2022). https://www.cnbc.com/2022/02/23/air-rage-during-the-pandemic-where-it-is-and-isnt-happening-.html.

5. Nick Rufford, "Snooze Control : Some Airlines Raise Temperature to Lull Passengers," *Los Angeles Times* (September 7, 1993). https://www.latimes.com/archives/la-xpm-1993-09-07-fi-32631-story.html.

6. Eyder Peralta, "Physicists Test for Most Efficient Way to Board a Plane," *NPR* (August 31, 2011). https://www.npr.org/sections/thetwo-way/2011/08/31/140097557/physicists-test-most-efficient-way-to-board-a-plane.

7. Dan Marzullo, "Gen Z Says Social Justice Is More Important Than Climate Change," Workest, March 18, 2021. https://www.zenefits.com/workest/gen-z-says-social-justice-is-more-important-than-climate-change/.

8. Nathaniel Meyersohn, "Why Nike Is Betting Its Slogan on Colin Kaepernick," *CNN* (September 30, 2018). https://www.cnn.com/2018/09/30/business/nike-colin-kaepernick-nfl-just-do-it/index.html.

9. Alexander Smith, "Pepsi Yanks Controversial New Ad Amid Backlash," *NBC* (April 5, 2017). https://www.nbcnews.com/news/nbcblk/pepsi-ad-kendall-jenner-echoes-black-lives-matter-sparks-anger-n742811.

10. Katherine A. DeCelles and Michael I. Norton, "Physical and Situational Inequality on Airplanes Predicts Air Rage," *Proceedings of the National Academy of Sciences of the United States of America* 113, no. 20 (May 17, 2016): 5588–5591. https://doi.org/10.1073/pnas.1521727113.

11. Mark Johanson, "What's Driving the US Air-Rage Spike?" *BBC Worklife* (June 29, 2021). https://www.bbc.com/worklife/article/20210629-whats-driving-the-us-air-rage-spike.

12. Mary Meisenzahl and Grace Dean, "From Toilet Paper to Candy Bars, Companies Hide Rising Costs by Shrinking the Size of Everyday Products. Here's What That Looks Like," *Business Insider* (August 25, 2022). https://www.businessinsider.com/shrinkflation-grocery-stores-pringles-cereal-candy-bars-chocolate-toilet-paper-cadbury-2021-7.

13. "Brands Take Note: Gen Z Is Putting Its Money Where Its Values Are," Sustainable Brands, May 8, 2018. https://sustainablebrands.com/read/walking-the-talk/brands-take-note-gen-z-is-putting-its-money-where-its-values-are.

14. Peter McIndoe, "The History," *Birds Aren't Real*, n.d. https://birdsarentreal.com/pages/the-history.

15. "The Origins of 'Birds Aren't Real,'" *60 Minutes*, CBS News (May 2, 2022). https://www.cbsnews.com/news/birds-arent-real-origin-60-minutes-2022-05-01/.

16. @birdsarentreal, "Ex-CIA Agent Eugene Price Confirms Everything," TikTok, November 2, 2020. https://www.tiktok.com/@birdsarentreal/video/6890737266478697733?lang=en.

17. "Birds Aren't Real at Twitter Headquarters," Instagram, November 12, 2021. https://www.instagram.com/p/CWMg0_Tv55T/.

18. Taylor Lorenz, "Birds Aren't Real, or Are They? Inside a Gen Z Conspiracy Theory," *New York Times* (December 9, 2021). https://www.nytimes.com/2021/12/09/technology/birds-arent-real-gen-z-misinformation.html.

19. Giovanni Luca Ciampaglia and Filippo Menczer, "Biases Make People Vulnerable to Misinformation Spread by Social Media," *Scientific American* (June 21, 2018). https://www.scientificamerican.com/article/biases-make-people-vulnerable-to-misinformation-spread-by-social-media/.

20. Ben Popken, "Age, Not Politics, Is Biggest Predictor of Who Shares Fake News on Facebook, Study Finds," *NBC News* (January 10, 2019). https://www.nbcnews.com/tech/tech-news/age-not-politics-predicts-who-shares-fake-news-facebook-study-n957246.

21. "10 Surprising Ways Boomers and Millennials Can Learn Anti-Phishing from Gen Z," *Phriendly Phishing* (blog), December 10, 2021. https://www.phriendlyphishing.com/blog/10-surprising-ways-boomers-and-millennials-can-learn-anti-phishing-from-gen-z.

Chapter 13

1. Dan Levin, "Generation Z: Who They Are, in Their Own Words," *New York Times* (March 28, 2019). https://www.nytimes.com/2019/03/28/us/gen-z-in-their-words.html?searchResultPosition=8.
2. Mr. Carrot's name has been changed for privacy.
3. "Happily Ever Avatar (TV Series 2018–)," IMDB, n.d. https://www.imdb.com/title/tt9212974/.
4. Rick Marshall, "Happily Ever Avatar: How HBO Max Series Found Love Connections in Gaming World," Digital Trends, July 2, 2020. https://www.digitaltrends.com/movies/happily-ever-avatar-hbo-max-love-relationships-in-gaming/.
5. "Avatar Life – Love, Metaverse App," Google Play Store, n.d. https://play.google.com/store/apps/details?id=com.xp101.ava_rus&pli=1
6. "Romance in the Clouds," Second Life Destinations, n.d. https://go.secondlife.com/destination/romance-in-the-clouds.
7. Mary Brune, "Gen Z Gamers: Key Insights," Newzoo, August 31, 2021. https://newzoo.com/insights/articles/gen-z-gamers-key-insights.
8. Dr. Justin Lehmiller, "Gen Z Aren't Having the Sex You Think: Here's Why," *Lovehoney* (blog), June 26, 2022. https://www.lovehoney.ca/blog/gen-z-are-having-less-sex-here-is-why.html.

Chapter 14

1. Nicholas Jones, Rachel Marks, Roberto Ramirez, and Merarys Rios-Vargas, "2020 Census Illuminates Racial and Ethnic Composition of the Country," United States Census Bureau,

August 12, 2021. https://www.census.gov/library/stories/2021/08/improved-race-ethnicity-measures-reveal-united-states-population-much-more-multiracial.html.

2. Nicholas Jones et al., "2020 Census Illuminates Racial and Ethnic Composition of the Country."

3. "3 Ways That the U.S. Population Will Change over the Next Decade," *PBS News Hour* (January 2, 2020). https://www.pbs.org/newshour/nation/3-ways-that-the-u-s-population-will-change-over-the-next-decade.

4. Marc Griffin, "Noname Pleads with Black Artists to Gatekeep the Culture," Yahoo! Life, January 20, 2023. https://www.yahoo.com/lifestyle/noname-pleads-black-artists-gatekeep-204714748.html.

5. "Bridging the Blues," Visit the USA, n.d. https://www.visittheusa.com/trip/bridging-blues.

6. "Blues: Definition, Artists, History, Characteristics, Types, Songs, & Facts," *Encyclopedia Britannica* (April 26, 2023). https://www.britannica.com/art/blues-music.

7. "W.C. Handy," Songwriters Hall of Fame, n.d. https://www.songhall.org/profile/WC_Handy.

8. Paul Resnikoff, "Led Zeppelin Openly Admitted to Plagiarism Multiple Times, Court Documents Reveal," Digital Music News, April 19, 2016. https://www.digitalmusicnews.com/2016/04/19/robert-plant-and-jimmy-page-blatantly-admit-to-stealing-their-music-led-zeppelin/.

9. Hugh Fielder, "The Led Zeppelin Songs That Led Zeppelin Didn't Write," Louder, May 13, 2016. https://www.loudersound.com/features/the-led-zeppelin-songs-that-led-zeppelin-didnt-write.

10. Gavin Edwards, "Led Zeppelin's 10 Boldest Rip-Offs," *Rolling Stone* (June 22, 2016). https://www.rollingstone.com/feature/led-zeppelins-10-boldest-rip-offs-223419/.

11. "Led Zeppelin's Whole Lotta Love Voted Best Guitar Riff," *BBC News* (August 25, 2014). https://www.bbc.com/news/entertainment-arts-28929167.

12. Rebecca Jennings, "Renegade TikTok Dance by Jalaiah Harmon Opens up Questions about Who Owns a Viral Dance," *Vox*

(February   4,   2020).   https://www.vox.com/the-goods/2020/2/4/21112444/renegade-tiktok-song-dance.

13. Global.Jones, "Arkansas Boy," TikTok, October 5, 2019. https://www.tiktok.com/@global.jones/video/6744451303356861701.

14. "Addison Rae," Wikipedia, accessed April 22, 2023. https://en.wikipedia.org/wiki/Addison_Rae.

15. Michelle Felix, ""They Make Millions off of It": Charli D'Amelio Publicly Criticized by Sunny Hostin for 'Stealing' TikTok Dances," Sportskeeda, June 29, 2021. https://www.sportskeeda.com/pop-culture/news-they-make-millions-it-charli-d-amelio-publicly-criticized-sunny-hostin-stealing-tiktok-dances.

16. Taylor Lorenz, "Meet the Original Renegade Dance Creator: Jalaiah Harmon," *New York Times* (August 28, 2021). https://www.nytimes.com/2020/02/13/style/the-original-renegade.html.

17. Cache McClay, "Why Black TikTok Creators Have Gone on Strike," *BBC News* (July 15, 2021). https://www.bbc.com/news/world-us-canada-57841055.

18. Cache McClay, "Why Black TikTok Creators Have Gone on Strike."

19. Jason Parham, "A People's History of Black Twitter, Part I," *WIRED* (July 15, 2021). https://www.wired.com/story/black-twitter-oral-history-part-i-coming-together/.

20. Patrick Thibodeau, "After Self-Criticism, Workday Improves Black Representation," Tech Target, July 28, 2021. https://www.techtarget.com/searchhrsoftware/news/252504556/After-self-criticism-Workday-improves-Black-representation#:~:text=A%202018%20study%20by%20the%20Center%20for%20Employment,software%20market%2C%20also%20falls%20below%20the%20UMass%20benchmark.

21. Kevin Dolan, Dame Vivian Hunt, Sara Prince, and Sandra Sancier-Sultan. "Diversity Still Matters," McKinsey & Co., May 19, 2020. https://www.mckinsey.com/featured-insights/diversity-and-inclusion/diversity-still-matters.

22. Kevin Dolan et al., "Diversity Still Matters."

23. Taylor Lorenz, "Meet the Original Renegade Dance Creator: Jalaiah Harmon," *New York Times* (August 28, 2021). https:// www.nytimes.com/2020/02/13/style/the-original-renegade .html.

Chapter 15

1. "Gen Alpha & Gen Z – The Future of Gaming: Newzoo Gamer Insights Report," Newzoo, September 28, 2022. https://newzoo .com/resources/trend-reports/gen-alpha-gen-z-the-future-of-gaming-free-report.

2. "Roblox Squid Game (Last Man Standing Wins 1 Million Robux)," FGTeeV Channel, YouTube, October 4, 2021. https:// www.youtube.com/watch?v=K_9TsVKYPp8.

3. Lucas Hill-Paul, "Squid Game Creator Defends Series from Hunger Games Comparison Backlash 'My Own Work,'" Express.Co.UK, October 20, 2021. https://www.express.co.uk/ showbiz/tv-radio/1498129/Squid-Game-showrunner-hits-back-Hunger-Games-comparisons.

4. "Rust Game Warning Label," Steam Gaming Store, n.d. https:// store.steampowered.com/agecheck/app/252490/.

5. Bruno Bettelheim, 2010. *The Uses of Enchantment: The Meaning and Importance of Fairy Tales*. (New York: Vintage Books), p. 45. First published 1976.

6. Bruno Bettelheim, 2010. *The Uses of Enchantment: The Meaning and Importance of Fairy Tales*. (New York: Vintage Books), p. 57. First published 1976.

Chapter 16

1. "George Santos for Congress," n.d., https://georgeforny .com/about/.

2. Grace Ashford and Michael Gold, "New York Republican George Santos's Résumé Called Into Question," *New York Times* (December 19, 2022). https://www.nytimes.com/2022/12/ 19/nyregion/george-santos-ny-republicans.html.

3. "SEC Charges Harbor City Capital with Securities Fraud," Silver Law Group, *Securities Arbitration Lawyers Blog*, May 5, 2021. https://www.silverlaw.com/blog/sec-charges-another-florida-internet-marketing-company-with-securities-fraud-harbor-city-capital/.

4. George Santos, "Will we landlords ever be able to take back possession of our property?" Twitter, February 8, 2021. https://twitter.com/Santos4Congress/status/1358981815495704581?ref_src=twsrc%5Etfw%7Ctwcamp%5Etweetembed%7Ctwterm%5E1358981815495704581%7Ctwgr%5E3e44837163aa5b60a220914417b58f61f4ef40ac%7Ctwcon%5Es1_&ref_url=https%3A%2F%2Fwww.newsweek.com%2Ffull-list-george-santos-claims-that-have-now-been-debunked-1770172.

5. Netflix. "This Whole Story Is Completely True," Twitter, February 22, 2022. https://twitter.com/netflix/status/1492234486498660354?s=20.

6. Yessi Bello Perez, "Meet the Dad Who Registered His Daughter's Birth on the Blockchain," Coindesk, September 11, 2021. Accessed May 3, 2023. https://www.coindesk.com/markets/2015/11/14/meet-the-dad-who-registered-his-daughters-birth-on-the-blockchain/.

7. "E-Democracy & Open Data," E-Estonia, November 23, 2022. https://e-estonia.com/solutions/e-governance/e-democracy/.

8. "2022 Corruption Perceptions Index," Transparency International, January 31, 2023. https://www.transparency.org/en/cpi/2022.

9. Savannah Fortis, "California County Approves Blockchain-Based Digital Wallet for Gov't Services," Cointelegraph, April 27, 2023. https://cointelegraph.com/news/california-approves-blockchain-based-digital-wallet-for-gov-t-services?ck_subscriber_id=2010635610.

10. "The Texting Generation: How Millenials and Gen-Z Prefer Text Over Face-to-Face Meetings," Womans Club of Carlsbad, June 14, 2022. https://womansclubofcarlsbad-com.ngontinh24.com/article/the-texting-generation-how-millenials-and-gen-z-prefer-text-over-face-to-face-meetings.

11. Scott Pelley, "Facebook Whistleblower Frances Haugen Details Company's Misleading Efforts on 60 Minutes," *CBS News* (October 4, 2021). https://www.cbsnews.com/news/facebook-whistleblower-frances-haugen-misinformation-public-60-minutes-2021-10-03/.

12. "Rule 6: Ask Yourself If It Sparks Joy," KonMari, the Official Website of Marie Kondo, November 3, 2021. https://konmari .com/marie-kondo-rules-of-tidying-sparks-joy/.

13. "Experience Is Everything: Here's How to Get It Right," PwC, 2018. https://www.pwc.com/us/en/services/consulting/library/consumer-intelligence-series/future-of-customer-experience.html.

14. "Generation Influence: Gen Z Study Reveals a New Digital Paradigm," *Business Wire* (July 7, 2020). https://www.business wire.com/news/home/20200706005543/en/Generation-Influence-Gen-Z-Study-Reveals-a-New-Digital-Paradigm.

15. "Explained: What Is Moon Farming, and Why Is It Popular?" CNBCTV18.Com, January 26, 2022. https://www.cnbctv18 .com/cryptocurrency/explained-what-is-moon-farming-and-why-is-it-popular-12253422.htm.

Chapter 17

1. Kyle Smith, "Working from Home Is No Longer Fun – It's Time to Go Back to the Office," *New York Post* (June 5, 2021). https://nypost.com/2021/06/05/why-its-time-to-end-work-from-home-and-go-back-to-the-office/.

2. Natalie Wong, Jeremy C.F. Lin, Paul Murray, and Noah Buhayar, "Remote Work Is Killing Manhattan's Commercial Real Estate Market," *Bloomberg* (September 25, 2022). https://www.bloomberg.com/graphics/2022-remote-work-is-killing-manhattan-commercial-real-estate-market/.

3. Tarab Zaidi, "British Billionaire Entrepreneur Thinks Letting People Work from Home Is 'Staggeringly Self-Defeating,'" *Business Today* (December 13, 2022). https://www.businessto day.in/latest/world/story/british-billionaire-entrepreneur-

thinks-letting-people-work-from-home-is-staggeringly-self-defeating-356227-2022-12-13.

4. John Caudwell, "'A Catastrophe for the British Economy': Phones 4U Founder John Caudwell Condemns WFH Culture," *Mail Online* (May 15, 2022). https://www.dailymail.co.uk/debate/article-10818091/A-catastrophe-British-economy-Phones-4u-founder-JOHN-CAUDWELL-condemns-WFH-culture.html.

5. André Spicer, "What Jacob Rees-Mogg, Alan Sugar and the Daily Mail Get Wrong About Home Working," *Guardian* (December 21, 2022). https://www.theguardian.com/commentisfree/2022/dec/21/jacob-rees-mogg-alan-sugar-daily-mail-home-working-woke.

6. J. David Goodman, "These Are the Perks Companies Use to Get Workers Back to Their Offices," *New York Times* (October 30, 2020). https://www.nytimes.com/2020/10/30/nyregion/new-york-city-office-coronavirus.html.

7. Chandni Kazi and Claire Hastwell. "Remote Work Productivity Study Finds Surprising Reality: 2-Year Analysis," *Great Place to Work* (blog), February 10, 2021. https://www.greatplacetowork.com/resources/blog/remote-work-productivity-study-finds-surprising-reality-2-year-study.

8. Jack Kelly, "How CEOs and Workers Feel About Working Remotely or Returning to the Office," *Forbes* (March 19, 2021). https://www.forbes.com/sites/jackkelly/2021/03/19/how-ceos-and-workers-feel-about-working-remotely-or-returning-to-the-office/?sh=37301a7529d9.

9. "Americans Are Embracing Flexible Work—and They Want More of It," McKinsey & Co., June 23, 2022. https://www.mckinsey.com/industries/real-estate/our-insights/americans-are-embracing-flexible-work-and-they-want-more-of-it.

10. "Employees Willing to Make Less Money to Stay Home," *Blind Blog – Workplace Insights*, April 5, 2021. https://www.teamblind.com/blog/index.php/2021/04/05/employees-willing-to-make-less-money-to-stay-home/.

11. J. David Goodman, "These Are the Perks Companies Use."

12. J. David Goodman, "These Are the Perks Companies Use."

13. Anne, "Stanford Study Shows Working from Home Is More Productive," *VSee* (blog), November 21, 2011. https://vsee .com/blog/stanford-study-shows-working-from-home-is-more-productive/.

14. Elena Cavender, "Instagram Says Gen Z Will Embrace the Social Media Side Hustle in 2023," *Mashable* (December 8, 2022). https://mashable.com/article/instagram-gen-z-monetize-social-media.

15. Chloe Taylor, "Major Trading Firm Lifts Ban on Side Hustles – Opening the Door for Employees to Become YouTubers or Startup Founders," *Fortune* (February 6, 2023). https://fortune .com/2023/02/06/mitsui-co-lifts-ban-on-second-job-side-hustles/.

16. Steve Mollman, "ChatGPT and Its Ilk Are Making It Easier for Remote Workers to Secretly Hold Two or More Full-Time Jobs," *Fortune* (April 21, 2023). https://fortune.com/2023/04/15/ai-chatgpt-remote-workers-overemployed-multiple-full-time-jobs/.

17. "Leveling the Playing Field in the Hybrid Workplace," *Future Forum Pulse Report*, January 2022. https://futureforum.com/wp-content/uploads/2022/01/Future-Forum-Pulse-Report-January-2022.pdf.

18. Ruchika Tulshyan, "Return to Office? Some Women of Color Aren't Ready," *New York Times* (July 23, 2021). https://www .nytimes.com/2021/06/23/us/return-to-office-anxiety.html.

19. Hugh Son and Dawn Giel, "Jamie Dimon, Fed Up with Zoom Calls and Remote Work, Says Commuting to Offices Will Make a Comeback," *CNBC* (May 4, 2021). https://www.cnbc .com/2021/05/04/jamie-dimon-fed-up-with-zoom-calls-and-remote-work-says-commuting-to-offices-will-make-a-comeback.html.

20. Joshua Meyer, "Gordon Gekko's Famous 'Greed Is Good' Line Was Inspired by an Actual Wall Street Criminal," *Film* (January 29, 2023). https://www.slashfilm.com/1176683/gordon-gekkos-famous-greed-is-good-line-was-inspired-by-an-actual-wall-street-criminal/.

21. Donald Liebenson, "How the *Glengarry Glen Ross* 'Coffee Is for Closers' Scene Got Made," *Vanity Fair* (October 4, 2022). https://www.vanityfair.com/hollywood/2022/10/how-the-glengarry-glen-ross-coffee-is-for-closers-scene-got-made.

22. Shawn Achor, Andrew Reece, Gabriella Rosen Kellerman, and Alexi Robichaux, "9 Out of 10 People Are Willing to Earn Less Money to Do More-Meaningful Work," *Harvard Business Review* (November 6, 2018), https://hbr.org/2018/11/9-out-of-10-people-are-willing-to-earn-less-money-to-do-more-meaningful-work.

23. B. Bastian, J. Jetten, and L. J. Ferris, "Pain as Social Glue: Shared Pain Increases Cooperation," *Psychological Science* (November 2014). DOI: 10.1177/0956797614545886.

24. "List of *The Backyardigans* Characters," *Nick Jr.* Wiki, n.d. https://nickjr.fandom.com/wiki/List_of_The_Backyardigans_characters.

25. Karl Moore, "MIT's Ed Schein on Why Corporate Culture Is No Longer the Relevant Topic and What Is," *Forbes* (November 29, 2011). https://www.forbes.com/sites/karlmoore/2011/11/29/mits-ed-schein-on-why-corporate-culture-in-no-longer-the-relevant-topic-and-what-is/?sh=4dc9734a5078.

26. Marcel Schwantes, "Steve Jobs Once Gave Some Brilliant Management Advice on Hiring Top People," *Inc.*, n.d. https://www.inc.com/marcel-schwantes/this-classic-quote-from-steve-jobs-about-hiring-employees-describes-what-great-leadership-looks-like.html.

27. Chandni Kazi and Claire Hastwell, "Remote Work Productivity Study Finds Surprising Reality: 2-Year Analysis," *Great Place to Work* (blog), February 10, 2021. https://www.greatplacetowork.com/resources/blog/remote-work-productivity-study-finds-surprising-reality-2-year-study.

Chapter 18

1. Bruno Bettelheim, 2010. *The Uses of Enchantment: The Meaning and Importance of Fairy Tales* (New York: Vintage Books), p. 73. First published 1976.

Epilogue

1. "Gardening in the Ancient World: Persian Gardens," *History of Garden Design and Gardening*, chapter 1, n.d. https://www.gardenvisit.com/book/history_of_garden_design_and_gardening/chapter_1_gardening_in_the_ancient_world/persian_gardens.
2. Kevin Kelly, "Everything That Doesn't Work Yet" *The Technium* (blog), n.d. https://kk.org/thetechnium/everything-that/.
3. Natalie Campisi, "Millions of Americans Are Still Missing Out on Broadband Access and Leaving Money on the Table – Here's Why," *Forbes Advisor* (October 28, 2022). https://www.forbes.com/advisor/personal-finance/millions-lack-broadband-access/.
4. Jane Thier, "Gen Z Exposes a Shortcoming of Their College Education That Could Impact the Workforce for Decades," *Fortune* (February 16, 2023). https://fortune.com/2023/02/16/gen-z-skills-gap-careers-college/?utm_source=email&utm_medium=newsletter&utm_campaign=data-sheet&utm_content=2023021719pm&tpcc=nldatasheet.
5. Media.Monks, "Making the Metaverse Accessible and Inclusive by Design," n.d. https://media.monks.com/articles/making-metaverse-accessible-and-inclusive-design.

Acknowledgments

WORDS FEEL SO flimsy at times like these! To my esteemed colleagues Michael Barngrover, Domingo Aguilar, Riaad v.d. Merwe, Asahi Ruiz, and Alex Dankloff for their inspiration; to Carlos Bernal for his magic work with Midjourney on the cover art; and my dear friends JR Vonk, Justin Bolognino, Rebecca Meijlink, Raoul Marques, Stacey Benjamin, and Anita Sanders, for their love and patience; to Brearley girls, through truth and toil, and my darling family, for their love and support.

—CDH

With love and thanks to my very own Virtual Natives, my sons, Otso and Ahti. I can only hope that as a parent I've taught you at least a fraction of everything that you've both taught me. And to my husband, Ismo, the most magnificent and supportive partner ever. Thanks for joining us on this wild ride!

—LS

About the Authors

Catherine D. Henry is a globally renowned expert on the Metaverse and Web3, SVP at Media.Monks and a strategic adviser to boards of directors and executive management teams, shaping how the world's most successful companies will enter, market, and monetize this new media landscape. With an MA in economics and MBA in marketing, Catherine brings over 20 years of experience in investment banking as an institutional investment adviser on frontier technologies. She is a thought leader on global tech megatrends, author of a series of Web3 and Metaverse Intelligence Reports, has lived and worked on four continents, and speaks five languages. She is African American.

Leslie Shannon is a Silicon Valley–based futurist and corporate adviser who focuses on connectivity-related tech disruptions and opportunities, including developments in robotics, drones, visual analytics, cloud gaming, and especially augmented and virtual reality – the foundations of the metaverse. Leslie has a BA from the University of Virginia, a master's degree from Yale University, and was a five-time champion on *Jeopardy!* She is the author of the book *Interconnected Realities*, a study of the purpose of the metaverse, and does all her daily fitness work in virtual reality.

Index